OXFORD STATISTICAL SCIENCE SERIES

Optimum Experimental Designs

A. C. ATKINSON

London School of Economics

and

A. N. DONEV

University of Kent, Canterbury

CLARENDON PRESS · OXFORD

This book has been printed digitally in order to ensure its continuing availability

OXFORD
UNIVERSITY PRESS

Great Clarendon Street, Oxford OX2 6DP

Oxford University Press is a department of the University of Oxford.
It furthers the University's objective of excellence in research, scholarship,
and education by publishing worldwide in

Oxford New York

Auckland Bangkok Buenos Aires Cape Town Chennai
Dar es Salaam Delhi Hong Kong Istanbul Karachi Kolkata
Kuala Lumpur Madrid Melbourne Mexico City Mumbai Nairobi
São Paulo Shanghai Singapore Taipei Tokyo Toronto
with an associated company in Berlin

Oxford is a registered trade mark of Oxford University Press
in the UK and in certain other countries

Published in the United States
by Oxford University Press Inc., New York

A catalogue record for this book is available from the British Library

Library of Congress Cataloging in Publication Data
Atkinson, A. C. (Anthony Curtis)
Optimum experimental designs/A. C. Atkinson and A. N. Donev.
(Oxford statistical science series; 8)
1. Experimental design. 2. Mathematical optimization. I. Donev,
A. N. (Alexander N.) II. Title. III. Series.
QA279.A82 1992 519.5—dc20 92-8342

ISBN 0-19-852254-1 (Hbk)

For

Ruth and Lubov

'. . . amidst the alien corn.'

Preface

A well-designed experiment is an efficient method of learning about the world. Because experiments in the world, and even in laboratories, cannot avoid random error, statistical methods are essential for their efficient design and analysis.

The fundamental idea behind this book is the importance of the model relating the reponses observed in an experiment to the experimental factors. The purpose of the experiment is to find out about the model, including its adequacy. The model can be very general: one receiving attention in several chapters is that for response surfaces in which the response is a smoothly varying function of the settings of the experimental variables. Experiments can then be designed to answer a variety of questions about the model. Often interest is in obtaining estimates of the parameters and using the fitted model for prediction. The variances of parameter estimates and predictions depend upon the experimental design and should be as small as possible. The unnecessarily large variances and imprecise predictions resulting from a poorly designed experiment waste resources.

The tool that we use to design experiments is the theory of optimum experimental design. The great power of this theory is that it leads to algorithms for the construction of designs which, as we show, can be applied in a wide range of circumstances. One purpose of the book is to describe enough of the theory to make apparent the overall pattern. However, as the title of the book makes clear, the emphasis is on the designs themselves. We are concerned both with their properties and with methods for their construction. The Appendix includes a Fortran program for the construction of designs, including many described in the examples.

The material has been divided into two main parts. The first eight chapters, 'Fundamentals', discuss the advantages of the statistical approach to the design of experiments and introduce many of the models and examples which are used in later chapters. The examples are, in the main, drawn from science and engineering. The same principles are applicable to agricultural experimentation, but the methods of design construction described here are not the most efficient for agricultural field trials. However, whatever the area of experimentation, the ideas of Part I are fundamental. These include an introduction to the ideas of models and least squares fitting. The ideas of optimum experimental design are introduced through the comparison of the variances of parameter estimates and the variance of the predicted response from a variety of designs and models. In Part II the relationship between these two set of variances leads to the General Equivalence Theorem which, in turn, leads to algorithms for designs. As well as these ideas, Part I includes, in Chapter 7, a description of standard designs. In order to keep the book to a reasonable length, there is rather little material on the analysis of experimental results. However, Part I concludes, in Chapter 8, with some examples of such analyses both numerical and graphical.

Part II opens with a discussion of the general theory of optimum design, followed by a discussion of a variety of criteria that may be appropriate for designing an experiment. Of these the most often used is D-optimality, which is the subject of Chapter 11. Succeeding chapters describe experiments with mixtures and designs for extensions to response surface models to include qualitative factors. Examples include the batch of a raw material or the particular design of a chemical reactor in addition to quantitative factors such as time and temperature. Chapter 15 is concerned with algorithms for the construction of designs, one of which is implemented in the Appendix.

Each chapter is intended to cover a self-contained topic. As a result, the chapters are of varying lengths. One of the shortest is Chapter 16 which covers designs for arbitrary design regions: despite the shortness of the chapter, the ability to calculate good designs under such non-standard conditions is a strong argument in favour of the application of optimum design theory. Chapter 18 describes the extension of the methods to non-linear regression models. The resulting designs depend upon prior estimates of the parameter values. Chapter 19 describes ways of incorporating uncertain prior knowledge into designs. Chapters 20 and 21 extend the discussion to multi-purpose designs and to those for discrimination between models.

In the last chapter we gather together a number of important topics. These include designs for off-line quality control (often known as Taguchi methods) which provide systematic methods for developing products which behave well under a broad range of conditions of use. Related ideas are used to design products which are insensitive to manufacturing fluctuations. This can be thought of as extending the statistical techniques of off-line quality control to the prevention, rather than mere removal, of substandard product. Other topics included in this chapter are designs for sequential clinical trials, for generalized linear models, and for computer simulation experiments.

It is hard to overrate the importance of the design of experiments. Sloppily designed experiments not ony waste resources, but may completely fail to provide answers. This is particularly true in the study of systems with several interacting factors. But, in a competitive world, an experiment which takes too long to develop an improved product is also useless: the market will have moved on before the results can be applied. Despite the importance of statistical methods for the design of experiments, the subject is seriously neglected in the training of many statisticians, scientists, and technologists. Obviously, in writing this book we have had in mind students and practitioners of statistics. But there is also much here of importance for anyone who has to perform experiments in the laboratory or factory. So in writing we also had in mind experimenters, the statisticians who sometimes advise them, and anyone who will be training scientists and technologists in universities, technical colleges, or on industrial short courses. The material of Part I, which is at a relatively low mathematical level, should be accessible to members of all these groups. The mathematical level of Part II is slightly higher, but we have avoided derivations of mathematical results—these can be found in the references and suggestions for further reading at the ends of most chapters. Little previous statistical knowledge is assumed. Although a first course in statistics with an introduction to regression would be helpful for Part I, such knowledge is not essential.

The idea of writing the book together came to us when we were both at Imperial College, London. We almost immediately put as many geographical obstacles in the way of completion as we could. Anthony Atkinson first went to Minneapolis and then moved to the London School of Economics, while Alexander Donev returned to Bulgaria in time for the momentous events of 1989. It seems churlish to complain about the effect of such historical occurrences on the postal service, but it is fair to say that they were not helpful. We are very grateful to the Bulgarian Ministry of Education and to the Staff Research Fund of the London School of Economics, both of which provided funds for us to meet and continue our collaboration.

One of the rewards of an active research interest is the friendships that it generates. We have benefited particularly, over the years, from discussions on the design of experiments with Rosemary Bailey, Kathryn Chaloner, David Cox, Valery Fedorov, and Henry Wynn. At the London School of Economics, Martin Knott and Neil Shephard enthusiastically provided a most necessary service in making a motley collection of computers, printers, and packages provide us with the desired hard copy. We are grateful to Dagmar Schumacher who processed the text for many of the chapters. Her patience and orderliness were exemplary. Jane Pugh drew many of the pictures and all the best ones. Samantha Firth considered mutiny about the amount of photocopying involved in communication between two authors in different places, but, in the event, remained cheerful and helpful. Finally, we note with regret the appreciable sexual stereotyping implied by this list.

London and Canterbury   A.C.A.
July 1991   A.N.D.

Contents

Part I Fundamentals

Part II Theory and applications

Part I
Fundamentals

1
Introduction

1.1 Some examples

This book is concerned with the design of experiments when uncontrollable fluctuations are appreciable compared with the effects to be investigated. Statistical methods are essential for experiments which provide unambiguous answers with a minimum of effort and expense, particularly if the effects of several factors are to be studied. The emphasis is on designs derived using the theory of optimum experimental design. Two main contributions result from such methods. One is the provision of algorithms for the construction of designs, and the other is the availability of quantitative methods for the comparison of proposed experiments. As we shall see, many of the widely used standard designs are optimum in ways to be defined in later chapters. The algorithms of the theory allow the incorporation of non-standard features into designs, often with little loss of efficiency.

The book is in two parts. In the first the ideas of the statistical design of experiments are introduced. In Part II, beginning with Chapter 9, the theory of optimum experimental design is devloped and numerous applications are described. To begin we describe three examples of experimental design. Comments on the scope and limits of the statistical contribution to experimental design are given in the second section. The chapter concludes with a guide to the literature.

Example 1.1 The desorption of carbon monoxide

During the nineteenth century the gas works, in which coal was converted to coke and town gas, was a major source of chemicals and fuel. At the end of the twentieth century, as the reality of a future with reduced supplies of oil approaches, the gasification of coal is again being studied. Typical of this renewed interest is the series of experiments described by Sams and Shadman (1986) on the potassium-catalysed production of carbon monoxide from carbon dioxide and carbon.

In the experiment, graphitized carbon was impregnated with potassium carbonate. This material was then heated in a stream of 15 per cent carbon dioxide in nitrogen. The full experiment used two complicated temperature–time profiles and several responses were measured. The part of the experiment of interest here consisted in measuring the total amount of carbon monoxide desorbed. These results are given in Table 1.1, together with the initial potassium/carbon (K/C) ratio, and are plotted in Fig. 1.1. They show a clear, seemingly linear, relationship between the carbon monoxide desorbed and the initial K/C ratio.

Some experimental design questions raised by this experiment are as follows.

1. Six levels of K/C ratio were used. Why six levels, and why these six?

2. The numbers of replications at the different K/C ratios vary from 2 to 6. Again why?

The purpose of questions such as these is to find out whether it is possible to do better by a different selection of numbers of levels and replicates. Better, in this context, means obtaining more precise answers for less effort. We shall be concerned with the need to define the questions which an experiment such as this is intended to answer. Once these are established, designs can be compared for answering a variety of questions and efficient experimental designs can be found.   □

Example 1.1 is an experiment with one continuous factor. In principle, an experimental run or trial could be performed for any non-negative value of the K/C ratio, although there will always be limits imposed by technical considerations, such as the strength of the apparatus, as well as by the values of the factors that are of interest to the experimenter. But often the factors in an experiment are qualitative, having a few fixed levels.

***Example* 1.2** Viscosity of elastomer blends

Derringer (1974) reports the results of an experiment on the viscosity of styrene butadiene rubber (SBR) blends. He introduces the experiment as follows:

'Most commercial elastomer formulations contain various amounts and types of fillers and/or plasticizers, all of which exert major effects on the viscosity of the system. A means of predicting the viscosity of a proposed formulation is obviously highly desirable since viscosity control is crucial to processing operations. To date, considerable work has been done on the viscosity of elastomer–filler systems, considerably less on elastomer–plasticizer systems, and virtually none on the complete elastomer–filler–plasticizer systems. The purpose of this work was the development of a viscosity model for the elastomer–filler–plasticizer system which could be used for prediction'.

Some of his experimental results are given in Table 1.2. The response is the viscosity of the elastomer blend. There are two continuously variable quantitative factors, the levels of filler and of naphthenic oils, both of which are measured in parts per hundred (phr) of the pure elastomer. The single qualitative factor is the kind of filler, of which there are three. The factor is therefore at three levels.

Table 1.1. Example 1.1: the desorption of carbon
monoxide

Observation number	Initial K/C atomic ratio (%)	CO desorbed (mole/mole C) (%)
1	0.05	0.05
2	0.05	0.10
3	0.25	0.25
4	0.25	0.35
5	0.50	0.75
6	0.50	0.85
7	0.50	0.95
8	1.25	1.42
9	1.25	1.75
10	1.25	1.82
11	1.25	1.95
12	1.25	2.45
13	2.10	3.05
14	2.10	3.19
15	2.10	3.25
16	2.10	3.43
17	2.10	3.50
18	2.10	3.93
19	2.50	3.75
20	2.50	3.93
21	2.50	3.99
22	2.50	4.07

For each of the three fillers, the experimental design is a 4×6 factorial, i.e. measurements of viscosity are taken at all combinations of four levels of naphthenic oil and of the six levels of filler. The questions that this design raises are extensions of those asked about Example 1.1.

1. Why four equally spaced levels of naphthenic oil?
2. Why six equally spaced levels of filler?
3. In order to investigate filler–plasticizer systems it is necessary to vary both factors together: experiments in one which factor is varied while the other is held constant will fail unless the two factors act independently. But is a complete factorial with 24 trials necessary? If the viscosity varies smoothly with the levels of the two factors, a simple polynomial model will explain the relationship. As we shall see in Part II, we do not need so many design points for such models.

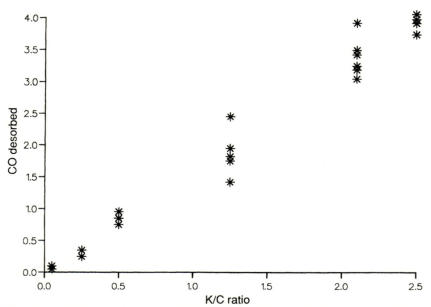

Fig. 1.1. Example 1.1: the desorption of carbon monoxide. Yield (carbon monoxide desorbed) against K/C ratio.

4. If there is a common structure for the different fillers, can this be used to provide an improved design?

The purpose of asking such questions is to focus on finding experimental designs which provide sufficiently accurate answers with a minimum of trials. To extend this example slightly, if there were two factors at four levels and two at six, the complete factorial design would require $24 \times 24 = 576$ trials. The use of a fractional factorial design, or other optimum design, would lead either to saving of money or to answering more questions with the same experimental effort.   □

These two examples are both taken from the technological literature and are typical of the applications lying behind the theory and designs we shall develop. We shall be rather less concerned with the kind of experiment which typically arises in agriculture where there is emphasis on design for the reduction of experimental error.

Example **1.3** Breaking strength of cotton fibres

Cox (1958, p. 26) discusses an agricultural example taken from Cochran and Cox (1957, §4.23). The data, given in Table 1.3, are from an experiment in which five fertilizer treatments $T_1, \ldots, T_5$ are applied to cotton plants. The

Table 1.2. Example 1.2: viscosity of elastomer blends (Mooney viscosity MS_4 at 100 °C as a function of filler and oil levels in SBR-1500)

Naphthenic oil (phr)	Filler	Filler level (phr)					
		0	12	24	36	48	60
0	A	26	28	30	32	34	37
	B	26	38	50	76	108	157
	C	25	30	35	40	50	60
10	A	18	19	20	21	24	24
	B	17	26	37	53	83	124
	C	18	21	24	28	33	41
20	A	12	14	14	16	17	17
	B	13	20	27	37	57	87
	C	13	15	17	20	24	29
30	A	—	12	12	13	14	14
	B	—	15	22	27	41	63
	C	11	14	15	17	18	25

The fillers are as follows: A, N990, Cabot Corporation; B, Silica A, Hi-Sil 223, PPG Industries; C, Silica B, Hil-Sil EP, PPG Industries.

response is the breaking strength of the cotton fibres. (Fertilizer helps the plants to grow, but what about the quality of the product?)

In such agricultural experiments there is often great variation in the yields on experimental plots of land even when they receive the same treatment combination, in this case fertilizer. In order to reduce the effect of this variation, which can mask the presence of treatment effects, the experimental plots are gathered into blocks: plots in the same block are expected to have more in common than plots in different blocks. A block might consist of plots from a single field or from one farm. In Table 1.3 there are three blocks of five plots each. A different treatment is given to each plot within the block. The primary interest in the experiment is the differences between the yields for the various treatments. The differences between blocks are usually of lesser importance. □

Randomized block experiments such as this, often elaborated to allow for several blocking factors, are described in most books on the design of experiments. References are given in the suggestions for further reading at the end of the chapter. Since these designs, with all factors qualitative, have been widely studied, described, and employed, we shall not repeat this material here. However, there is one feature of this experiment which makes it an example of a class of problems with which we are concerned.

Table 1.3. Example 1.3: breaking strength of cotton fibres

	T_1	T_2	T_3	T_4	T_5	Total	Mean
	Treatments						
Block 1	7.62	8.14	7.76	7.17	7.46	38.15	7.63
Block 2	8.00	8.15	7.73	7.57	7.68	39.13	7.83
Block 3	7.93	7.87	7.74	7.80	7.21	38.55	7.71
Total	23.55	24.16	23.23	22.54	22.35	115.83	7.72
Mean	7.85	8.05	7.74	7.51	7.45	7.72	

The description has been in terms of five fertilizer treatments $T_1, \ldots, T_5$. However, these treatments are five levels of a continuous factor, the amount of potash per acre. We then have an experimental design with one quantitative factor, the levels of which can be chosen in the presence of a qualitative factor, representing the three blocks. A difference between this experiment and Example 1.2 is that here the block effects can be viewed as nuisance parameters, whereas the differences between polymers in Example 1.2 were of equal interest with the effects of the additives.

1.2 Scope and limitations

The common structure to all examples is the allocation of treatments, or factor combinations, to experimental units. In Example 1.2 the unit would be a specimen of pure elastomer which is then blended with specified amounts of filler and naphthenic acid, which are the treatments. In Example 1.3 the unit is the plot of land receiving a unique treatment combination, here a level of fertilizer.

In the optimum design of experiments the allocation of treatments to units depends upon the model or models that are to be used to explain the data and the questions that are asked about the models. In Example 1.1 the question might be whether the relationship between carbon monoxide and potassium carbonate was linear, or whether some curvature was present, requiring a second-order term in the model. The optimum design for answering this question is different from that for fitting a first-order model. The theory could be used to find a design efficient for detection of curvature. Another possibility is to find designs efficient for a specified set of purposes, with weightings attached to the relative importance of the various aspects of the design.

However, there are aspects of the design which are not of direct statistical concern. The purpose of the experiment and the design of the apparatus are outside the scope of statistics, as are the size of the units and the responses to be measured. In Example 1.1 the amount of carbon needed for a single experiment at a specified carbonate level depends upon the design of the apparatus. Likewise, the experimental region depends upon the knowledge and intentions of the experimenter. Once the region has been defined, the techniques decribed in this book are concerned with the choice of treatment combinations, which may be from specified levels of qualitative factors, or values from ranges of quantitative variables. The total size of the experiment will depend upon the resources, both of money and time, which are available. The statistical contribution at this level is often to calculate the size of effects which can be detected, with reasonable certainty, in the presence of errors of the size anticipated in the particular experiment.

This is a book about an important, powerful, and very general method for the design of experiments. The product of the algorithms and tables is a list of treatment to be applied to the experimental units. Which units receive which treatments must be arranged in such a way as to avoid systematic bias. Usually this is achieved by randomizing the application of treatments to units, to avoid the confounding of treatment effects with those due to omitted variables which are nevertheless of importance. In many technological experiments time of day is an important factor if an apparatus is switched off overnight. Randomization of treatment allocation over time provides insurance against the confounding of observed treatment effects with time of day.

Many books on the design of experiments contain almost as much material on the analysis of data as on the design of experiments. We focus on design. Analysis is mentioned specifically only in Chapter 8. In the first part of the book, of which Chapter 8 is the last chapter, we give the background to our approach. Chapters 9 onwards are concerned with the central theory of optimum experimental design, illustrated with numerous examples.

1.3 Background reading

There is a vast statistical literature on the design of experiments. Cox (1958) remains a relatively short non-mathematical introduction to the basic ideas. Many books are primarily concerned with the experimental designs which are of special interest in agriculture where, as in Example 1.3, treatments are relatively unstructured and the variability between units is high. John (1971) gives a good account of designs of this kind, together with factorial designs and their fractions. His book is typical of many in that it does not mention optimum experimental design. Similar material is covered, for example, by John and Quenouille (1977) and Mead (1988), which emphasizes agricultural experiments. A more applied approach is provided by Pearce (1983), which distils a lifetime's experience of agricultural experimentation. Box *et al.* (1978) provide a stimulating introduction to statistics and experimental design which reflects the authors' experience in the chemical industry. More advanced topics in design, especially for response surfaces, i.e. designs when all factors are quantitative, are covered by Box and Draper (1987). The papers assembled in Ghosh (1990) cover many topics in the design of industrial experiments.

The pioneering book, in English, on optimum experimental design is Fedorov (1972). Silvey (1980) provides a concise introduction to the central theory of the General Equivalence Theorem. From a mathematical viewpoint much of our book can be considered as a series of special cases of the theorem, which we give in Chapter 9. The book of Pazman (1986) is, in contrast, resolutely algebraic. Shah and Sinha (1989) is concerned solely with designs for qualitative factors.

There are several books in German on optimum design. Bandemer *et al.* (1973) provides a brief introduction, at the opposite extreme from the two-volume handbook of Bandemer *et al.* (1977) and Bandemer and Näther (1980). Other books from the former German Democratic Republic include Rasch and Herrendörfer (1982) and, in English translation, Bunke and Bunke (1986). Although this last is more concerned with analysis than with design, Chapter 8 does give a summary of optimum experimental design. Russian books include Ermakov (1983) and Ermakov and Zhigliavsky (1987).

The optimum design of experiments is based on a theory which, like any general theory, provides a unification of many separate results and a way of generating new results in novel situations. Part of this flexibility results from the algorithms derived from the General Equivalence Theorem, combined with computer-intensive search methods. However, there are other approaches to experimental design, particularly for some response surface problems, which lead to appealing and widely used designs. Box and Draper (1987, Chapter 14) give a criticism of optimum experimental design starting from the premise that all models are only an approximation to the truth.

The algorithms of optimum design theory are not, in general, the best way of constructing designs in which the factors are either all qualitative, or quantitative but with specified levels to be used a specified number of times. For such problems combinatorial methods of construction, as described by Street and Street (1987), are preferable. The discrete mathematics underlying this approach is presented by Biggs (1989). Finally, brief mention has been made of the desirability of randomization, a topic given authoritative coverage by Bailey (1991), who defines many reasons for the randomization of experiments.

2
Some key ideas

2.1 Scaled variables

Figure 2.1 is one schematic representation of an experiment. A single trial consists of measuring the values of the t response, or output, variables $y_1, \ldots, y_t$. These values are believed to depend upon the value of the m factors or explanatory variables $u_1, \ldots, u_m$. However, the relationship is obscured by the presence of unobservable random errors $\varepsilon_1, \ldots, \varepsilon_t$.

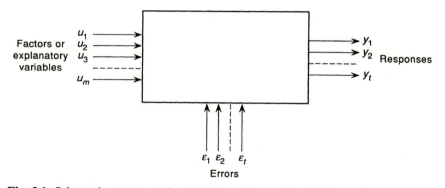

Fig. 2.1. Schematic representation of an experiment. The relationship between the factors u and the response y is obscured by the presence of error ε. The values of u are to be chosen by the experimenter who observes y but not ε.

Quantitative factors, such as the K/C ratio in Example 1.1, can take any value in a specified interval. Such factors are often called predictors or explanatory variables. Typically they vary between a minimum and a maximum value so that

$$u_{i, \min} \leqslant u_i \leqslant u_{i, \max} \qquad (i = 1, \ldots, m). \qquad (2.1)$$

The values of the upper and lower limits $u_{i, \max}$ and $u_{i, \min}$ depend upon the physical limitations of the system and upon the range of the factors thought by the experimenter to be interesting. For example, if pressure is one of the factors, the experimental range will be bounded by the maximum safe working pressure of the apparatus. However, $u_{i, \max}$ may be less than this value if such high pressures are not of interest.

It is convenient for most applications to scale the quantitative variables. The unscaled variable $u_1, \ldots, u_m$ are replaced by standardized, or coded, variables, which are often, but not invariably, scaled to lie between -1 and 1. For such a range the coded values are defined by

$$x_i = \frac{u_i - u_{i0}}{\Delta_i} \qquad (i = 1, \ldots, m) \qquad (2.2)$$

where

$$u_{i0} = u_{i, \min} + \frac{u_{i, \max} - u_{i, \min}}{2}$$

and

$$\Delta_i = u_{i, \max} - u_{i0} = u_{i0} - u_{i, \min}.$$

Although designs will be decribed in terms of the coded variables, it is sometimes desirable to return to the original values of the factors, particularly for the interpretation of the experimental results. The reverse transformation to (2.2) yields

$$u_i = u_{i0} + x_i \Delta_i \qquad (i = 1, \ldots, m).$$

2.2 Design regions

If the limits (2.1) apply independently to each of the m factors, the experimental region in terms of the scaled factors x_i will be an m-dimensional cube. For $m = 2$ this is the square shown in Fig. 2.2(a).

The cubic design region is the most frequently encountered for quantitative variables. However, the nature of the experiment may sometimes cause more complicated specification of the factor intervals and of the design region. For example, the region will be spherical if it is defined by the equation

$$\sum_{i=1}^{m} x_i^2 \leqslant R^2$$

where the radius of the sphere is R. This design region for $m = 2$ is shown in Fig. 2.2(b). Such a region suggests equal interest in departures in any direction from the centre of the sphere, which might be current experimental or operational conditions.

We shall also give some consideration to mixture experiments in which the response depends only on the proportions of the components of a mixture and not at all on the total amount. One example is the octane rating of a petrol (gasoline) blend. An important feature of such experiments is that a change in

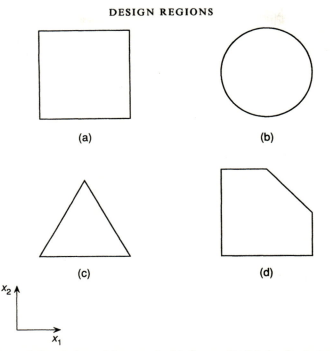

Fig. 2.2. Some design regions: (a) square (cubic for $m > 2$); (b) circular (spherical); (c) simplex, for mixture experiments; (d) restricted, to avoid simultaneous high values of x_1 and x_2.

the level of one of the factors leads to a change in the values of one, some, or all of the other factors. The constraints

$$\sum_{i=1}^{q} x_i = 1 \qquad x_i \geqslant 0$$

imposed on the q mixture components make the design region a $(q-1)$ dimensional simplex. Figure 2.2(c) shows a design region for a three-component mixture.

In addition to quantitative factors, we shall also consider experiments with qualitative factors, such as the type of filler in Example 1.2, which can take only a specified number of levels. Other examples are the sex of a patient in a clinical trial and the type of reactor employed in a chemical experiment. Qualitative factors are often represented in designs by indicator or dummy variables. An instance is Example 5.3 of §5.2.

Many experiments involve both qualitative and quantitative factors, as did Examples 1.2 and 1.3, the latter after some reinterpretation. Such experiments are the subject of Chapter 13. In addition, some of the quantitative factors might be mixture variables. The experimental regions may also be more complicated than those shown in Fig. 2.2, often because of the imposition of extra constraints. An example with which many readers will be familiar from their school-days is the ability of reactions in organic chemistry to produce tar, rather than the desired product, when run for a long time at a high temperature. Such areas of the experimental region are best avoided, leaving a region of the shape shown in Fig. 2.2(d). Designs for such restricted experimental regions are the subject of Chapter 16. Whatever the shape of the design region, which we call $\mathscr{X}$, the principles of experimental design remain the same. The algorithms of optimum design lead to a search over $\mathscr{X}$ for a design minimizing a function which often depends on the variances of the parameter estimates. The structure of $\mathscr{X}$ partially determines whether a standard design can be used or whether, and what kind of, a search is needed for the optimum design.

2.3 Random error

The observations y_i obtained at the N design points in the region $\mathscr{X}$ are subject to error. This may be of two kinds.

1. Systematic errors, or biases, due, for example, to an incorrectly calibrated apparatus. Such biases must, of course, be avoided. In part they come under the heading of the non-statistical aspects of the experiment discussed in §1.2. But randomization should be used to guard against other biases such as those mentioned in §1.2, which arise when treatments for comparisons are applied to units which differ in a systematic way.

2. Random errors. These are represented by the vector ε in Fig. 2.1. In the simplest case suppose that we have repeat measurements for which the model is

$$y_i = \mu + \varepsilon_i \qquad (i = 1, \ldots, 5). \tag{2.3}$$

Usually we shall assume additive and independent errors of constant variance (see Chapter 5). Let the five readings of the yield of a chemical reaction be

$$y_1 = 0.91, \ y_2 = 0.92, \ y_3 = 0.90, \ y_4 = 0.93, \text{ and } y_5 = 0.92.$$

The results, obtained under supposedly identical conditions, show random fluctuation. To estimate μ the sample mean $\bar{y}$ is used, where

$$\bar{y} = \frac{1}{n} \sum_{i=1}^{n} y_i.$$

The variance σ^2 of the readings is estimated by

$$s^2 = \frac{1}{n-1} \sum (y_i - \bar{y})^2.$$

For the five readings given above, $\bar{y} = 0.916$ and $s^2 = 0.0114$. The estimators will be different for different samples from the same system. For larger sample sizes the fluctuations in the estimates will be smaller.

Often the aims of the experiment will be to find an approximating function or model relating y_i to the m factors u_i. Usually the estimation of the variance σ^2 will be of secondary importance. Even so, a knowledge of σ^2 is important to provide a measure of random variability against which to assess observed effects. Formal methods of assessment include confidence intervals for parameter estimates and predictions of the response, and significance tests. These provide a mechanism for determining which terms should be included in a model.

Model (2.3) contains one unknown parameter μ. In general, there will be p parameters. In order to estimate these p parameters at least $N = p$ trials will be required at distinct points in the design region. However, if σ^2 is not known, but has to be estimated from the data, more trials will be required. The larger the residual degrees of freedom $v = N - p$, the better the estimate of σ^2, provided that the postulated model holds. If lack of fit of the model is a possibility of interest, σ^2 is better estimated from replicate observations. The analysis can then provide an estimate of both σ^2 and that portion of the residual sum of squares attributable to lack of fit. The analysis in Chapter 8 of the data from Table 1.1 provides an example.

3
Experimental strategies

3.1 Objectives of the experiment

We are mainly concerned with experiments where the purpose is to elucidate the behaviour of a system by fitting an approximating function or model. The distinction is with experiments where the prime interest is in estimating differences, or other contrasts, in yield between units receiving separate treatments. Often the approximating function will be a low-order polynomial. But, as in Chapter 18, the models may sometimes be non-linear functions representing knowledge of the mechanism of the system under study. There are several advantages to summarizing and interpreting the results of an experiment through a fitted model.

1. A prediction can be given of the responses under investigation at any point within the design region. Confidence intervals can be used to express the uncertainty in these predictions by providing a range of plausible values.

2. We can find the values of the factors for which the optimum value (maximum or minimum) of each response occurs. Depending upon the model, the values are found by either numerical or analytical optimization. The set of optimum conditions for each response is then a point in factor space, not necessarily one at which the response was measured during the experiment. Optimization of the fitted model may sometimes lead to estimated optimum conditions outside the experimental region. Such extrapolations are liable to be unreliable and further experiments are needed to check whether the model still holds in this new region of factor space.

3. When there are several responses, it may be desired to find a set of factor levels which ensure optimum, or near optimum, values of all responses. If, as is usually the case, the optima do not coincide, a compromise needs to be found. One technique is to weight the responses to reflect their relative importance and then to optimize the weighted combination of the responses.

4. A final advantage of the fitted model is that it allows graphical representation of the relationships being investigated. However, the conclusions of any analysis depend strongly on the quality of the fitted models and hence on the way in which the experiment is designed and carried out.

These general ideas are described in a specific context in the next example, which also illustrates the use of the scaled variables introduced in §2.1.

Example **3.1** The purification of nickel sulphate

The purpose of the experiment was to optimize the purification of nickel sulphate solution, the impurities being iron, copper, and zinc, all in the bivalent state. Petkova *et al.* (1987) investigate the effect of five factors on six responses. Table 3.1 gives the maximum and minimum values of the unscaled factors u_i and the corresponding coded values x_i.

Table 3.1. Example 3.1: the purification of nickel sulphate. The five factors and their uncoded and coded values

	Uncoded value u_i		Coded value x_i	
Factor	Minimum	Maximum	Minimum	Maximum
1 Time of treatment (min)	60	120	−1	1
2 Temperature (°C)	65	85	−1	1
3 Consumption of $CaCO_3$ (%)	100	200	−1	1
4 Concentration of zinc (g/dm^3)	0.1	0.4	−1	1
5 Mole ratio Fe/Cu	0.91	1.39	−1	1

Since iron, copper, and zinc are impurities, high deposition of these three elements was required. These are given as the first three responses Y_1, Y_2, and Y_3 in Table 3.2. Low loss of nickel was also important, and is denoted by Y_4. Two further responses are Y_5, the ratio of the final concentration of nickel and zinc, and Y_6, the pH of the final solution. Target values were specified for all six responses.

From previous experience it was expected that second-order polynomials would adequately describe the response. The experimental design, given in Table 3.2, consists of a 2^{5-1} fractional factorial plus star points, a form of composite design discussed in §7.5 for investigating second-order models. Table 3.2 also gives the observed values of the six responses. The 26 trials of the experiment were run in random order, not in the standard order of the table.

The analysis consisted of fitting a separate second-order model for each response. Contour plots of the fitted responses against pairs of important variables indicated appropriate experimental conditions for each response, from which an area in the design region was found in which all the responses seemed to satisfy the experimental requirements. These conditions were $u_1 = 90$ min, $u_2 = 80$ °C, and $u_4 = 0.175$ g/dm^3, with u_3 free to vary between 160 and 185 per cent and the ratio u_5 lying between 0.99 and 1.24. Further

Table 3.2. Example 3.1: the purification of nickel sulphate

Factors					Responses					
x_1	x_2	x_3	x_4	x_5	y_1	y_2	y_3	y_4	y_5	y_6
+1	+1	+1	+1	+1	94.62	99.98	99.83	9.19	104889	5.07
−1	+1	+1	+1	−1	100.00	99.97	95.12	7.57	3462	4.94
+1	−1	+1	+1	−1	100.00	99.99	93.81	7.68	2730	5.39
−1	−1	+1	+1	+1	77.01	99.99	91.39	6.69	2084	5.05
+1	+1	−1	+1	−1	89.96	82.63	24.58	1.38	239	2.62
−1	+1	−1	+1	+1	81.89	86.97	99.78	3.27	85988	2.90
+1	−1	−1	+1	+1	79.64	93.82	78.13	1.95	862	3.70
−1	−1	−1	+1	−1	88.79	85.53	7.04	1.06	195	3.10
+1	+1	+1	−1	−1	100.00	100.00	99.33	7.26	51098	4.92
−1	+1	+1	−1	+1	93.23	99.93	97.99	7.38	22178	5.07
+1	−1	+1	−1	+1	89.61	99.99	98.36	5.76	27666	5.29
−1	−1	+1	−1	−1	99.95	99.92	91.35	5.01	4070	5.02
+1	+1	−1	−1	+1	95.80	86.81	30.25	5.12	681	2.59
−1	+1	−1	−1	−1	86.59	83.99	38.96	1.30	599	2.66
+1	−1	−1	−1	−1	88.46	85.49	42.48	2.03	630	3.16
−1	−1	−1	−1	+1	70.86	91.30	28.45	1.25	663	3.20
−1	0	0	0	0	97.68	99.97	95.02	5.68	4394	4.80
+1	0	0	0	0	99.92	99.99	97.57	6.68	10130	5.18
0	−1	0	0	0	99.33	99.98	97.06	6.21	8413	5.08
0	+1	0	0	0	99.38	99.90	96.83	7.05	7748	4.90
0	0	−1	0	0	80.10	87.74	19.55	1.10	324	3.08
0	0	+1	0	0	98.57	99.98	99.31	6.40	35655	5.28
0	0	0	−1	0	98.64	99.95	97.09	6.00	28239	4.80
0	0	0	+1	0	99.07	100.00	93.94	6.54	3169	4.81
0	0	0	0	−1	99.96	99.95	82.55	5.52	1910	5.04
0	0	0	0	+1	97.68	100.00	89.06	6.30	3097	5.13

The five responses are as follows: $y_1 - y_3$, deposition of iron, copper, and zinc; y_4, loss of nickel; y_5, ratio of final concentration of nickel to final concentration of iron; y_6, pH of the final solution. The experiment is given in standard order.

experimentation under these conditions confirmed that the values of all six responses were satisfactory. □

3.2 Stages in experimental research

The experiment described in §3.1 is one in which a great deal was known a priori. This information included the following.

1. The five factors known to affect the responses.

2. Suitable experimental ranges for each factor.

3. An appropriate model, in this case a second-order polynomial for each response.

An appreciable part of this book is concerned with designs for second-order models, which are appropriate in the region of a maximum or minimum of the response. However, it may be that the results of the experiment allow a simpler representation, with few or no second-order terms. Examples of model simplification are discussed in Chapter 8.

Such refinements of models usually occur far along the experimental path of iteration between experimentation and model-building. In this section we discuss some of the earlier stages of experimental programmes. A typical sequence of experiments leading to a design such as that of Table 3.2 is discussed in the next section.

1 *Background to the experiment.* The successful design of an experiment requires the evaluation and use of all prior information, even if only in an informal manner. What is known should be summarized and questions to be answered must be clearly stated. Factors which may affect the responses should be listed. The responses should both contain important information about the system and be measurable.

Although such strictures may seem platitudinous, the discipline involved in thinking through the purpose of the experimental programme is most valuable. If the programme involves collaboration, time used in clarifying knowledge and objectives is always well spent and often, in itself, highly informative.

2 *The choice of factors.* At the beginning of an experimental programme there will be many factors which, separately or jointly, may have an effect on the response. Some initial effort is often spent in screening out those factors that matter from those that do not. First-order designs, such as those of §22.3 and the 2^{6-3} fractional factorial of Table 7.3, are suitable for this stage. Second-order designs, such as the composite design of Table 3.2, are used at a later stage when trials are made near to a minimum or maximum of the response.

It is assumed that quantitative factors can be set exactly to any value in the design region, independently of one another. At the same time factors which will not be varied during the whole experiment will remain unchanged. An exception is in the design of mixture experiments, such as those of Chapter 12, where changing the amount of one component must change the amount of at least one other component.

The intervals over which quantitative factors are varied during the experiment need to be chosen with care. If they are too small, the effect of the factor on the response may be swamped by experimental error. On the other hand, if the intervals are too wide, the underlying relationship between the factors and the response can become too complicated to be respresented using a reasonably simple model.

3 *The reduction of error.* If there is appreciable variability between experimental units, these can be grouped together into blocks of more similar units, as in Example 1.3, with an appreciable gain in accuracy. Some examples of the division of 2^m factorials into blocks are given in §7.3. More general blocking of response surface designs is described in Chapter 14.

The choice of an appropriate blocking structure is often of great importance, especially in agricultural experiments, and can demand much skill from the experimenter. In technological experiments, batches of raw material are frequently an important source of variability, the effect of which can be reduced by taking batches as a blocking factor. An alternative approach is to use numerical values of nuisance variables at the design stage. If the purity of a raw material is its important characteristic, the experiment can be designed using the measured values on the various batches. Similarly, an experiment can be designed to behave well in the presence of a quadratic trend in time by including the quadratic terms in the model, or by seeking to find a design orthogonal to the trend.

4 *The choice of model.* Optimum experimental designs depend upon the model relating the response to the factors. The model needs to be sufficiently complicated to approximate the main features of the data, without being so complicated that unnecessary effort is involved in estimating a multitude of parameters. Some suitable models are described in the next chapter.

5 *Design criterion and size of the design.* The size of the design is constrained by resources, usually cost and time. The precision of parameter estimates increases with the number of trials, but also depends upon the location of the design points. Several design criteria are described in Chapter 10. These lead to designs maximizing information about specific aspects of the model.

6 *Choice of an experimental design.* In many cases the required experimental design can be found in the literature. Chapter 7 describes some standard designs. If a standard design is used, it is important that it takes into account all the features of the experiment, such as structure of the experimental region and the division of the units into blocks. If a standard design is not available, the methods of optimum experimental design will provide an appropriate design. In either case randomization in the application of treatments to units will be important.

7 *Conduct of the experiment.* The values of the responses should be measured for all trials. The measurements of the responses should be

statistically independent. If several observations are made over time on a single unit, the assumption of independence will be violated and the time series element in the data should be allowed for in the analysis.

If any values of the factors are set incorrectly, the actual values should be recorded and used in the analysis. If obtaining correct settings of the factors is likely to present difficulties, experimental designs should be used with only a few settings of each factor. Optimum experimental designs with each factor at only three or five levels, for example, can be found by searching over a grid of candidate points. The designs are frequently only slightly less efficient than those found by searching over a continuous region. The details are given in Chapter 15.

8 *Analysis of the data.* The results of the experiment can be summarized in tables such as those of Chapter 1. Very occasionally no further analysis is required. However, almost invariably, a preliminary graphical investigation will be informative, to be followed by a more formal analysis yielding parameter estimates and confidence intervals. Examples are given in Chapter 8.

Experimentation is iterative. The preceding list of points suggests a direct path from problem formulation to solution. However, at each stage the experimenter may have to reconsider decisions taken at earlier stages of the investigation. Problems arising at some stages may add to eventual understanding of the system.

If the model fitted at stage 8 appears inadequate, this may be because the model is too simple, or there may be errors and outliers in the data owing, for example, to failures in measurement and recording devices. In any case the design will need to be augmented and stages 6, 7, and 8 repeated until a satisfactory model is achieved.

3.3 The optimization of yield

Experiments for finding the conditions of maximum yield are often sequential and nicely illustrate the stages of an experimental programme. In the simplest case, described here, all factors are quantitative, with the response being a smooth function of the settings of the factors.

1 *Screening experiments.* First-order designs are often used to determine which of the many potential factors are important. The 2^{6-3} fractional factorial of Table 7.3 has already been mentioned. Other screening designs are described in §22.3.

2 *Initial first-order design.* As a result of the screening stage a few factors will emerge as being most important. In general let there be m. For illustration we take $m = 2$. The path of a typical experiment is represented in Fig. 3.1. The

initial design consists of a 2^m factorial, or perhaps a fraction if $m \geqslant 5$, with perhaps three centre points. If the average response at the centre is much the same as the average of the factorial points, the results can be represented by a first-order surface, i.e. a plane.

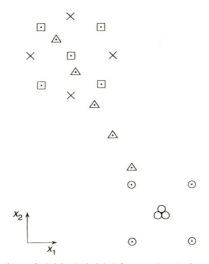

Fig. 3.1. The optimization of yield: $\odot$ initial first-order design with centre points; $\triangle$ path of steepest ascent; $\boxdot$ second first-order design; $\times$ star points providing a second-order design.

3 *Steepest ascent*. Experiments, illustrated by triangles in Fig. 3.1, are performed sequentially in the direction of steepest ascent which runs perpendicularly to the contours of the plane fitted in stage 2. Progress in this direction continues until the response starts to decrease, when the maximum in this direction will have been passed. In both this stage and the previous one assessments of differences between observed or fitted responses have to be made relative to the random error of the measurements.

4 *Second first-order design*. The squares in the figure represent a second first-order design, centred at or near the conditions of highest yield found so far. We suppose that, in this example, comparison of the average responses at the centre of the design and at the factorial points indicates curvature of the response surface. Therefore the experiment has taken place near a maximum, perhaps of the whole surface, but certainly in the direction of steepest ascent.

5 *Second-order design*. The addition of trials at other levels of the factors, the star points in Fig. 3.1, makes possible the estimation of second-order

terms. These may indicate a maximum in or near the experimental region. Or it may be necessary to follow a second path of steepest ascent along a ridge in the response surface orthogonal to the path of stage 3.

The formal description of the use of steepest ascent and second-order designs for the experimental attainment of optimum conditions was introduced by Box and Wilson (1951). Their work came from the chemical industry, where it is natural to think of the response as a yield which is to be maximized. In other situations the response might be a percentage of unacceptable product; for example, etched wafers in the production of chips in the electronics industry. Interest would then be in minimization of the response. However, the same experimental strategy is appropriate.

3.4 Further reading

Many statistical books on experimental design, especially Cox (1958) and Box *et al.* (1978), contain material on the purposes and strategy of experimentation. A different, although complementary, perspective on experimentation is provided by Wilson (1952), a chemist rather than a statistician. Mead (1988, especially §2.6) discusses the selection of blocks, particularly in agricultural field trials. Experiments in which observations are made over time on a single unit are often called repeated measures. For the analysis of such experiments see Crowder and Hand (1990). Chapter 15 of Box *et al.* (1978), 'Response surface methods', gives a fuller treatment of the material of §3.3.

4
The choice of a model

4.1 Linear models for one factor

Optimum experimental designs depend upon the model or models to be fitted to the data, although not usually, for linear models, on the values of the parameters of the models. This chapter is intended to give some advice on the choice of an appropriate form for a model. Whether or not the choice was correct can, of course, only be determined by analysis of the experimental results.

The true underlying relationship between the observed response y and the factors x is usually unknown. Therefore we choose as a model an approximating function which is likely to follow the response closely over the region of interest. Our concern will be mostly with polynomial models which are linear in the parameters. These can be thought of as Taylor series approximations to the true relationship. This section ends with some comments on non-linear models which can be linearized by transformation. Simple non-linear models are the subject of §4.2. We begin this section with the simplest linear model: the first-order model for a single factor.

Figure 4.1 shows the relationship between the expected response $\eta(x)$ and x for the three first-order models

$$\eta(x) = 16 + 7.5x \tag{4.1}$$

$$\eta(x) = 18 - 4x \tag{4.2}$$

$$\eta(x) = 12 + 5x. \tag{4.3}$$

These models describe monotonically increasing or decreasing functions of x in which the rate of increase of $\eta(x)$ does not depend on the value of x. The values of the two parameters determine the slope and intercept of each line. The use of least squares for estimating the parameters of such models once data have been collected is described in §5.1.

Increasing x in model (4.1) or model (4.3) causes $\eta(x)$ to increase without bound. Often, however, responses either increase to an asymptote with x, or even pass through a maximum and then decrease. Some suitable curves with a single maximum or minimum are shown in Fig. 4.2. The three models are

$$\eta(x) = 25 - 14x + 6x^2 \tag{4.4}$$

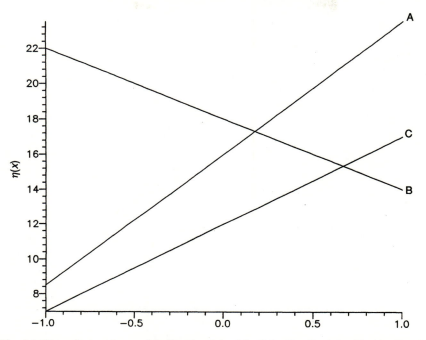

Fig. 4.1. Three first-order models: line A, $\eta(x) = 16 + 7.5x$; line B, $\eta(x) = 18 - 4x$; line C, $\eta(x) = 12 + 5x$.

$$\eta(x) = 20 - 10x + 40x^2 \tag{4.5}$$

$$\eta(x) = 50 + 5x - 35x^2. \tag{4.6}$$

A second-order, or quadratic, model is symmetrical about its extreme, be it a maximum or a minimum. For model (4.4) the minimum is at $x = 7/6$, outside the range of plotted values. For model (4.5), the extreme is a minimum, indicated by the positive coefficient (40) of x^2. For model (4.6), the maximum is at 1/14 and the coefficient of x^2 is negative. The larger the absolute value of the quadratic coefficient, the more sharply the single maximum or minimum is defined.

More complicated forms of response factor relationship can be described by third-order polynomials. Examples are shown in Fig. 4.3 for the models

$$\eta(x) = 90 - 85x + 16x^2 + 145x^3 \tag{4.7}$$

$$\eta(x) = 125 + 6x + 10x^2 - 80x^3 \tag{4.8}$$

$$\eta(x) = 62 - 25x + 70x^2 - 54x^3. \tag{4.9}$$

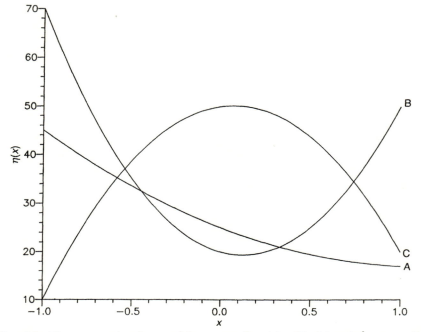

Fig. 4.2. Three second-order models: curve A, $\eta(x) = 25 - 14x + 6x^2$; curve B, $\eta(x) = 20 - 10x + 40x^2$; curve C, $\eta(x) = 50 + 5x - 35x^2$.

These three figures show that the more complicated the response relationship to be described, the higher the order of polynomial required. Although high-order polynomials can be used to describe quite simple relationships, the extra parameters will usually not be justified when experimental error is present in the observations to which the model is to be fitted. In general, the inclusion of unnecessary terms inflates the variance of predictions from the fitted model. Increasing the number of parameters in the model may also increase the size of the experiment, so providing an additional incentive for the use of simple models.

Experience indicates that in very many experiments the response can be described by polynomial models of order no greater than 2. Curves with multiple points of inflection, like those of Fig. 4.3, are rare. A more frequent form of departure from the models of Fig. 4.2 is caused by asymmetry around the single extreme point. This is often more succinctly modelled by a transformation of the factor, for example to $x^{1/2}$ or log x, than by the addition of a term in x^3. In this book we are mostly concerned with second-order models.

In addition to polynomial models, there is an important class of models with

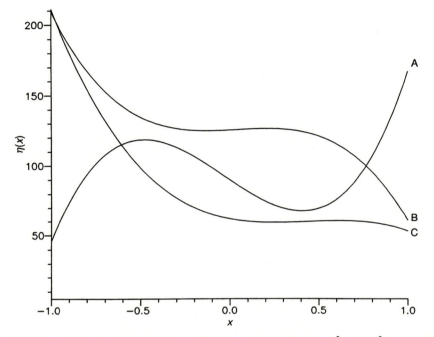

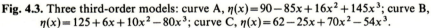

Fig. 4.3. Three third-order models: curve A, $\eta(x) = 90 - 85x + 16x^2 + 145x^3$; curve B, $\eta(x) = 125 + 6x + 10x^2 - 80x^3$; curve C, $\eta(x) = 62 - 25x + 70x^2 - 54x^3$.

non-linear parameters which, after transformation, can be regarded as linear. For example, the model

$$\eta(x) = \beta_0 x_1^{\beta_1} x_2^{\beta_2} \ldots x_m^{\beta_m} \qquad (4.10)$$

is non-linear in the parameters $\beta_1, \ldots, \beta_m$. Such models are used to describe the kinetics of chemical reactions and relationships between dimensionless quantities in technology and engineering. Taking logarithms of both sides of the model yields

$$\log \eta(x) = \log \beta_0 + \beta_1 \log x_1 + \cdots + \beta_m \log x_m, \qquad (4.11)$$

which can be written in the form

$$\tilde{\eta}(x) = \tilde{\beta}_0 + \beta_1 \tilde{x}_1 + \cdots + \beta_m \tilde{x}_m \qquad (4.12)$$

where

$$\tilde{\eta}(x) = \log \eta(x) \qquad \tilde{x}_j = \log x_j \qquad (j = 1, \ldots, m).$$

Thus (4.12) is a first-order polynomial in the transformed variables $\tilde{x}$. However, the equivalence of (4.10) and (4.12) in the presence of experimental error also requires that the errors in (4.10) are multiplicative so that they become additive in the transformed model (4.12).

4.2 Non-linear models

There are sometimes situations when models with non-linear parameters are to be preferred to attempts at linearization, particularly if the error assumptions are violated by the transformed model. Where the non-linear model arises from theory, estimation of the parameters will be of direct interest. In other cases the response surface can only be described succinctly by a non-linear model. As an example, the model

$$\eta(x) = \beta_0 \{1 - \exp(-\beta_1 x)\}$$

is plotted in Fig. 4.4 for $\beta_0 = 2.5$ and three values of β_1. In Fig. 4.4(a) $\beta_1 = 0.5$, in Fig. 4.4(b) $\beta_1 = 1.5$, and in Fig. 4.4(c) $\beta_1 = 4$. For all three sets of parameter values the asymptote has the same value of 2.5, but this value is approached more quickly as β_1 increases. Simple polynomial models of the type described in the previous section are not appropriate for models such as this which contain an asymptote.

One advantage of non-linear models is that they often contain few parameters when compared with polynomial models. A second advantage is that, if a non-linear model is firmly based in theory, extrapolation to values of x outside the region where data have been collected is unlikely to produce seriously misleading predictions. Unfortunately the same is not usually true for polynomial models. A disadvantage of non-linear models is that optimum designs for the estimation of parameters depend on the unknown values of the parameters. Designs for non-linear models are the subject of Chapter 18.

4.3 Interaction

The simplest extension of the polynomial models of §4.1 is to the first-order model in m factors. Figure 4.5 gives equispaced contours for the two-factor model

$$\eta(x) = 1 + 2x_1 + x_2.$$

The effects of the two factors are additive. Whatever the value of x_2, a unit increase in x_1 will cause an increase of 2 in $\eta(x)$.

Often, however, the factors interact, so that the effect of one factor depends upon the level of the other. Suppose that there are again two factors, one quantitative and the other, z, a qualitative factor at two levels, 0 and 1. Figure 4.6(a) shows a plot of $\eta(x)$ against x for the model

$$\eta(x) = 0.2 + 0.8x + 0.5z \qquad (z = 0, 1).$$

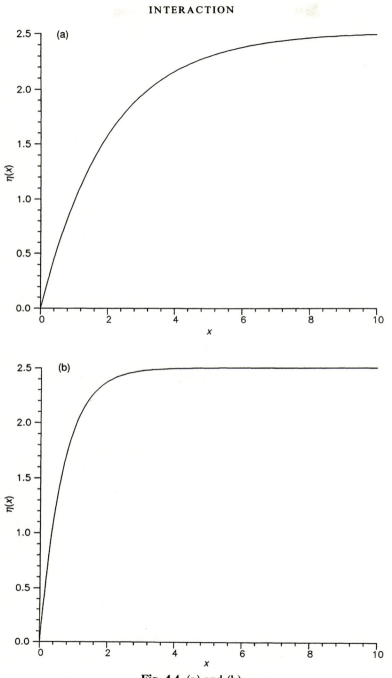

Fig. 4.4. (a) and (b).

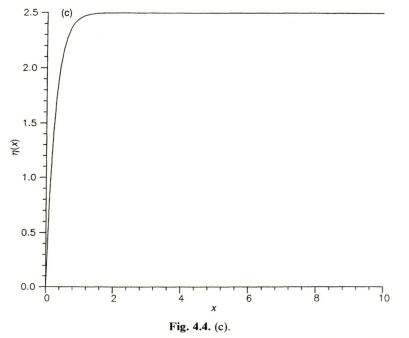

Fig. 4.4. (c).

A non-linear model with an asymptote: $\eta(x) = 2.5\{1 - \exp(\beta_1 x)\}$. (a) $\beta_1 = 0.5$; (b) $\beta_1 = 1.5$; (c) $\beta_1 = 4$.

For this model without interaction the effect of moving from the low to the high level of z is to increase $\eta(x)$ by 0.5. However, at either level, the rate of increase of $\eta(x)$ with x is the same. This is in contrast to Fig. 4.6(b) of the model

$$\eta(x) = 0.2 + 0.8x + 0.3z + 0.7xz \qquad (z = 0, 1).$$

The presence of the interaction term xz means that the rate of increase of $\eta(x)$ with x depends upon z. Instead of the two parallel lines of Fig. 4.6(a), Fig. 4.6(b) shows two straight lines which are not parallel. In the case of a strong interaction between x and z, the sign of the effect of x might even reverse between the two levels of z.

Interaction for two quantitative factors is illustrated in Fig. 4.7, where the model is

$$\eta(x) = 1 + 2x_1 + x_2 + 2x_1 x_2.$$

The effect of the interaction term $x_1 x_2$ is to replace the straight line contours of Fig. 4.5 by hyperbolae. For any fixed value of x_1, the effect of increasing x_2 is

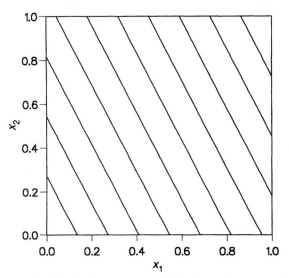

Fig. 4.5. Contours of $\eta(x) = 1 + 2x_1 + x_2$, a model with no interaction between x_1 and x_2.

constant, as can be seen by the equispaced contours. Similarly, the effect of x_1 is constant for fixed x_2. However, the effect of each variable depends on the level of the other.

Interactions frequently occur in the analysis of experimental data. An advantage of designed experiments in which all factors are varied to give a systematic exploration of combinations of factor levels is that interactions can be estimated. Designs in which one factor at a time is varied, all others being held constant, do not yield information about interactions. They will therefore be inefficient and uninformative if interactions are present. Interactions of third, or higher, order are possible, with the three-factor interaction involving terms like $x_1 x_2 x_3$. It is usually found that there are fewer two-factor interactions than significant main effects, and that higher-order interactions are proportionately less common. Pure interactions, i.e. interactions between factors the main effects of which are absent, are rare. One reason is that if the interaction is measured in terms of the scaled factors x, rewriting the interaction as a function of the unscaled factors u automatically introduces main effects of all factors present in the interaction.

4.4 Response surface models

Experiments in which all factors are quantitative frequently take place at or near the maximum or minimum of the response, i.e. in the neighbourhood of

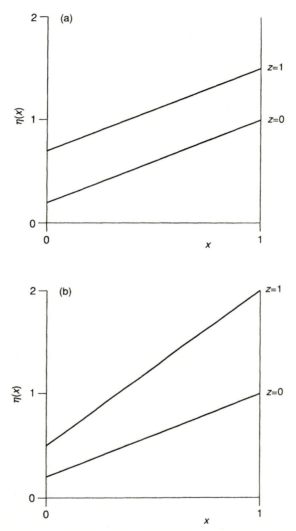

Fig. 4.6. Interaction between qualitative and quantitative factors: (a) no interaction, $\eta(x) = 0.2 + 0.8x + 0.5z$; (b) interaction, $\eta(x) = 0.2 + 0.8x + 0.3z + 0.7xz$.

conditions which are optimum according to some criterion. In order to model the curvature present, a full second-order model is required. Figure 4.8 shows contours of the response surface

$$\eta(x) = 1.27 - 2.9x_1 - 1.6x_2 + 2x_1^2 + x_2^2 + x_1 x_2$$
$$= 2(x_1 - 0.6)^2 + (x_2 - 0.5)^2 + (x_1 - 0.6)(x_2 - 0.5). \qquad (4.13)$$

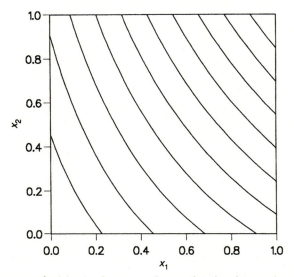

Fig. 4.7. Contours of $\eta(x) = 1 + 2x_1 + x_2 + 2x_1x_2$ showing interaction between two quantitative factors.

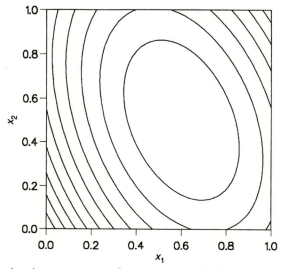

Fig. 4.8. Second-order response surface: contours of $\eta(x) = 1.27 - 2.9x_1 - 1.6x_2 + 2x_1^2 + x_2^2 + x_1x_2$.

The positive coefficients of x_1^2 and x_2^2 indicate that these elliptical contours are modelling a minimum. In the absence of the interaction term x_1x_2 the axes of the ellipse would lie along the co-ordinate axes. If, in addition, the coefficients of x_1^2 and x_2^2 are equal, the contours are circular. An advantage of writing the model in the form (4.13) is that it is clear that the ellipses are centred on $x_1 = 0.6$, $x_2 = 0.5$.

The second-order model (4.13) is of the same order in both factors. The ability to estimate all the coefficients of the model dictates the use of a second-order design such as the 3^2 factorial or the final design of Fig. 3.1. Often, however, the fitted model will not need all the terms and will be of different order in the various factors.

Example **4.1** Freeze drying

Table 4.1 gives the results of a 3^2 factorial experiment on the conditions of freeze drying (from Savova *et al.* 1989). The factors were the amount of glycerine x_1 per cent and the effect of the speed of freeze drying x_2 °C/min. The response y was the percentage of surviving treated biological material. The model which best describes the data is

$$\hat{y} = 90.66 - 0.5x_1 - 9x_2 - 1.5x_1x_2 - 3.5x_1^2, \tag{4.14}$$

which is second order in x_1, first order in x_2, and also includes an interaction term. $\qquad\qquad\square$

Table 4.1. Example 4.1: freeze drying. Percentage of surviving treated biological material

x_1 (amount of glycerine) (%)	x_2 (speed of freeze drying) (°C/min)		
	10	20	30
10	96	85	82
20	100	92	80
30	96	88	76

Since the response in Example 4.1 has a maximum value of 100 it might be preferable to fit a model which takes account of this bound, perhaps by using the transformation methods of §8.2. These may also lead to models of varying order in the factors. For example, the model with log y as a response fitted by Atkinson (1985, p. 131) to Brownlee's stack loss data (Brownlee 1965, p. 454) is exactly of the form of (4.14).

5
Models and least squares

5.1 Simple regression

The plots in Chapter 4 illustrate some of the forms of relationship between an m-dimensional factor x and the true response $\eta(x)$, which will often be written as $\eta(x, \beta)$, to stress dependence on a vector of p unknown parameters β. For a first-order model $p = m + 1$. Measurements of $\eta(x)$ are subject to error, giving observations y. Often the experimental error is additive and the model for the observations is

$$y_i = \eta(x_i, \beta) + \varepsilon_i \qquad (i = 1, \ldots, N). \tag{5.1}$$

If the error term is not additive, it is frequently possible to make it so by transformation of the response. For example, taking the logarithm of y makes a multiplicative error additive. Further discussion is given in Chapter 8 in the analysis of Example 1.2 (Derringer's elastomer data).

The absence of systematic errors implies that $E(\varepsilon_i) = 0$, where E stands for expectation. The customary second-order assumptions are

(i) $$E(\varepsilon_i \varepsilon_j) = \text{covar}(\varepsilon_i, \varepsilon_j) = 0 \qquad (i \neq j)$$

and

(ii) $$\text{var}(\varepsilon_i) = \sigma^2.$$

This assumption of independent errors of constant variance will invariably need to be checked. Violations of independence are most likely when the data form a series in time or space.

The second-order assumptions justify use of the method of least squares to estimate the vector parameter β. The least squares estimates $\hat{\beta}$ minimize the sum over all N observations of the sum of squared deviations

$$S(\beta) = \sum_{i=1}^{N} \{y_i - \eta(x_i, \beta)\}^2, \tag{5.2}$$

so that

$$S(\hat{\beta}) = \min_{\beta} S(\beta). \tag{5.3}$$

The formulation in (5.2) and (5.3) does not imply any specific structure for

$\eta(x, \beta)$. If the model is non-linear in some or all of the parameters β, the estimate $\hat{\beta}$ has to be found iteratively (see §18.4). However, if the model is linear in the parameters, explicit expressions can be found for $\hat{\beta}$.

The plot in Fig. 1.1 of the 22 readings on the desorption of carbon monoxide from Table 1.1 suggests that, over the experimental region, there is a straight line relationship between y and x of the form

$$\eta(x_i, \beta) = \beta_0 + \beta_1 x_i \qquad (i = 1, \ldots, N). \tag{5.4}$$

For this simple linear regression model

$$S(\beta) = \sum_{i=1}^{N} (y_i - \beta_0 - \beta_1 x_i)^2.$$

The minimum is found by differentiation, giving the pair of derivatives

$$\frac{\partial S}{\partial \beta_0} = -2 \sum_{i=1}^{N} (y_i - \beta_0 - \beta_1 x_i)$$

$$\frac{\partial S}{\partial \beta_1} = -2 \sum_{i=1}^{N} (y_i - \beta_0 - \beta_1 x_i) x_i. \tag{5.5}$$

At the minimum both derivatives are zero. Solution of the resulting simultaneous equations yields the least squares estimates

$$\hat{\beta}_1 = \frac{\Sigma y_i (x_i - \bar{x})}{\Sigma (x_i - \bar{x})^2}$$

with

$$\hat{\beta}_0 = \bar{y} - \hat{\beta}_1 \bar{x}, \tag{5.6}$$

where the sample averages are $\bar{x} = \Sigma x_i / N$ and $\bar{y} = \Sigma y_i / N$, and all summations are over $i = 1, \ldots, N$. Therefore the least squares line passes through $(\bar{x}, \bar{y})$ and has slope $\hat{\beta}_1$.

The distribution of $\hat{\beta}$ will depend upon the distribution of the errors ε_i. Augmentation of the second-order assumptions by the condition

(iii) the errors $\varepsilon_i \sim N(0, \sigma^2)$

yields the normal-theory assumptions. The parameter estimates $\hat{\beta}$, which, from (5.6), are linear combinations of normally distributed observations are then themselves normally distributed.

The least squares estimates are unbiased, i.e. $E(\hat{\beta}) = \beta$, provided that the correct model has been fitted. The variance of $\hat{\beta}_1$ is

$$\text{var}(\hat{\beta}_1) = \frac{\sigma^2}{\Sigma (x_i - \bar{x})^2}. \tag{5.7}$$

Usually σ^2 will have to be estimated, often from the residual sum of squares, giving the residual mean square estimate

$$s^2 = \frac{S(\hat{\beta})}{N-2} \qquad (5.8)$$

on $N-2$ degrees of freedom. The estimate will be too large if the model is incorrect. This effect of model inadequacy can be avoided by estimating σ^2 solely from replicated observations. An attractive feature of the design of Table 1.1 is that it provides a replication mean square estimate of σ^2 on 16 degrees of freedom. Another possibility is to use an external estimate of σ^2, based on experience, or derived from previous experiments. It is frequently found that such estimates are unrealistically small.

Whatever its source, let the estimate of σ^2 be s^2 on v degrees of freedom. Then to test the hypothesis that β_1 has the value β_{10},

$$\frac{\hat{\beta}_1 - \beta_{10}}{\{s^2/\Sigma(x_i - \bar{x})^2\}^{1/2}} \qquad (5.9)$$

is compared with the t distribution on v degrees of freedom. The $100(1-\alpha)$ per cent confidence limits for β are

$$\hat{\beta} \pm \frac{t_{v,\alpha}s}{\{\Sigma(x_i - \bar{x})^2\}^{1/2}}. \qquad (5.10)$$

The prediction from the fitted model at the point x, not necessarily included in the observations from which the parameters were estimated, is

$$\hat{y}(x) = \hat{\beta}_0 + \hat{\beta}_1 x = \bar{y} + \hat{\beta}_1(x - \bar{x}) \qquad (5.11)$$

with variance

$$\text{var}\{\hat{y}(x)\} = \sigma^2 \left\{ \frac{1}{N} + \frac{(x - \bar{x})^2}{\Sigma(x_i - \bar{x})^2} \right\}. \qquad (5.12)$$

The least squares residuals are defined to be

$$e_i = y_i - \hat{y}_i = y_i - \bar{y} - \hat{\beta}(x_i - \bar{x}). \qquad (5.13)$$

The use of the residuals in checking the model assumed for the data is exemplified in Chapter 8.

It is often convenient, particularly for more complicated models, to summarize the results of an analysis, including hypothesis tests such as (5.9), in an analysis of variance table. The decomposition

$$\Sigma(y_i - \bar{y})^2 = \Sigma(\hat{y}_i - \bar{y})^2 + \Sigma(y_i - \hat{y}_i)^2, \qquad (5.14)$$

where all summations are over the N observations, leads to Table 5.1. The entries in the column headed 'Mean square' are sums of squares divided by the

Table 5.1. Analysis of variance for simple regression

Source	Degrees of freedom	Sum of squares	Abbreviation	Mean square	F
Regression	1	$\Sigma(\hat{y}_i-\bar{y})^2$	SSR	SSR	SSR/s^2
Residual (error)	$n-2$	$\Sigma(y_i-\hat{y}_i)^2$	SSE	SSE/$(n-2)=s^2$	
Total (corrected)	$n-1$	$\Sigma(y_i-\bar{y})^2$	SST		

degrees of freedom. The F test for the regression is, in this case, the square of the t test (5.9). A numerical example of such a table is given in §8.1 as part of the analysis of the data on the desorption of carbon monoxide.

5.2 Matrices and experimental design

To extend the results of the previous section to linear models with $p>2$ parameters, it is convenient to use matrix algebra. The basic notation is established in this section.

The linear model will be written

$$E(Y)=F\beta \qquad (5.15)$$

where, in general, Y is the $N\times 1$ vector of responses, β is a vector of p unknown parameters, and F is the $N\times p$ extended design matrix. The ith row of F is $f^T(x_i)$, a known function of the m explanatory variables.

Example 5.1 Simple regression

For $N=3$, the simple linear regression model (5.4) is

$$E\begin{bmatrix} Y_1 \\ Y_2 \\ Y_3 \end{bmatrix} = \begin{bmatrix} 1 & x_1 \\ 1 & x_2 \\ 1 & x_3 \end{bmatrix} \begin{bmatrix} \beta_0 \\ \beta_1 \end{bmatrix}.$$

Here $m=1$ and $f^T(x_i)=(1 \; x_i)$.

In order to design the experiment it is necessary to specify the design matrix

$$X = \begin{bmatrix} x_1 \\ x_2 \\ x_3 \end{bmatrix}.$$

The entries of F are then determined by X and by the model. Suppose that the factor x is quantitative $-1\leqslant x\leqslant 1$. The design region is then written $\mathscr{X}=[-1,1]$. The design problem might be to choose N points in $\mathscr{X}$ so that the

Table 5.2. Designs and models for a single quantitative factor

Design	Design points x				Number of trials N
5.1	-1	0	1		3
5.2	-1	1	1		3
5.3	-1	$-1/3$	$1/3$	1	4

Models

First order, Example 5.1	$E(Y)=\beta_0+\beta_1 x$
Quadratic, Example 5.2	$E(Y)=\beta_0+\beta_1 x+\beta_2 x^2$
Quadratic with one qualitative factor,	$E(Y)=\alpha_j+\beta_1 x+\beta_2 x^2$
Example 5.3	$(j=1,\ldots,l)$

linear relationship between y and x given by (5.4) can be estimated as precisely as possible. One possible design for this purpose is Design 5.1 in Table 5.2, which consists of trials at three equally spaced values of X with design matrix

$$X_1 = \begin{bmatrix} -1 \\ 0 \\ 1 \end{bmatrix}.$$

Another possibility is Design 5.2, which has two trials at one end of the design region and one at the other. The design matrix is then

$$X_2 = \begin{bmatrix} -1 \\ 1 \\ 1 \end{bmatrix}. \qquad \square$$

Example **5.2** Quadratic regression

If the model is

$$E(Y_i)=\beta_0+\beta_1 x_i+\beta_2 x_i^2, \qquad (5.16)$$

allowing for curvature in the dependence of Y on x, trials will be needed at at least three different values of x in order to estimate the three parameters. The equally spaced four-trial Design 5.3, with design matrix

$$X_3 = \begin{bmatrix} -1 \\ -1/3 \\ 1/3 \\ 1 \end{bmatrix}.$$

would allow detection of departures from the quadratic model. For X_3 the extended design matrix for the quadratic model is

$$F = \begin{bmatrix} 1 & -1 & 1 \\ 1 & -1/3 & 1/9 \\ 1 & 1/3 & 1/9 \\ 1 & 1 & 1 \end{bmatrix},$$

where the final column gives the values of x_i^2.   □

Example 5.3 Quadratic regression with a single qualitative factor

The simple quadratic model (5.16) can be extended by assuming that the response depends not only on the quantitative variable x, but also on a qualitative factor z at l unordered levels. These might, for example, be l different designs of chemical reactor which are to be compared over a range of x values. Such models are the subject of Chapter 13.

Suppose that $l = 2$. If the effect of z is purely additive so that the response curve is moved up and down, as in Fig. 5.1, the model is

$$E(Y_i) = \alpha_j + \beta_1 x_i + \beta_2 x_i^2 \qquad (i = 1, \ldots, N; j = 1, 2) \qquad (5.17)$$

or, in matrix form,

$$E(Y) = W\gamma = E\alpha + F\beta. \qquad (5.18)$$

In general the matrix E in (5.18), of dimension $N \times l$, consists of indicator variables for the levels of z.

Suppose that the three-level Design 5.1 is repeated once at each level of z. Then

$$
\begin{array}{ccc}
E & : & F \\
\end{array}
$$

$$W = \begin{bmatrix} 1 & 0 & -1 & 1 \\ 1 & 0 & 0 & 0 \\ 1 & 0 & 1 & 1 \\ 0 & 1 & -1 & 1 \\ 0 & 1 & 0 & 0 \\ 0 & 1 & 1 & 1 \end{bmatrix}$$

The ith row of W is $w^T(x_i, z_i)$.

A more complicated model with a similar structure might be appropriate for the analysis of Derringer's data on the viscosity of elastomer blends (Example 1.2). Here, as there, one design question is whether the same design

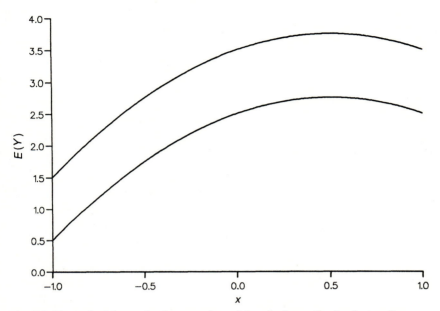

Fig. 5.1. Example 5.3: quadratic regression with a single qualitative factor. Response $E(Y)$ when the qualitative factor has two levels.

should be repeated at each level of z, if this is possible. If z is a blocking factor, the number of trials at each of the l levels may be specified, when the equal replication of this example may not be possible.

For $l=2$, other parameterizations are equivalent. For example the two distinct rows of E could be $(1\ -1)$ and $(1\ 1)$. The form used here is most appropriate for generalization to l levels.   □

5.3 Least squares

This section gives the extension of the least squares results of §5.1 to the linear model with p parameters (5.15). For this model the sum of squares to be minimized is

$$S(\beta) = (y - F\beta)^{\mathrm{T}}(y - F\beta). \tag{5.19}$$

The least squares estimates of β, found by differentiation of (5.19), satisfy the p least squares, or normal, equations

$$F^{\mathrm{T}}F\hat{\beta} = F^{\mathrm{T}}y. \tag{5.20}$$

The $p \times p$ matrix $F^{\mathrm{T}}F$ is the information matrix for β. The larger $F^{\mathrm{T}}F$, the greater is the information in the experiment. Experimental design criteria for comparing information matrices are discussed in Chapter 10.

By solving (5.20) the least squares estimates of the parameters are found to be

$$\hat{\beta} = (F^{\mathrm{T}}F)^{-1}F^{\mathrm{T}}y. \tag{5.21}$$

If the model is not of full rank, $F^{\mathrm{T}}F$ cannot be uniquely inverted, and only a set of linear combinations of the parameters can be estimated, perhaps a subset of β. In the majority of examples in this book, inversion of $F^{\mathrm{T}}F$ is not an issue. The covariance matrix of the least squares estimates is

$$\mathrm{var}\ \hat{\beta} = \sigma^2(F^{\mathrm{T}}F)^{-1}. \tag{5.22}$$

The variance of $\hat{\beta}_j$ is proportional to the jth diagonal element of $(F^{\mathrm{T}}F)^{-1}$: the covariance of $\hat{\beta}_j$ and $\hat{\beta}_k$ is proportional to the (j, k)th off-diagonal element. If interest is in the comparison of experimental designs, the value of σ^2 is not relevant, since the value is the same for all proposed designs for a specific experiment.

Tests of hypotheses about the individual parameters β_j can use the t test analogously to (5.9), with the variance of $\hat{\beta}_j$ from (5.22) in the denominator. For tests about several parameters, the F test is used. The related $100(1-\alpha)$ per cent confidence region for all p elements of β is of the form

$$(\beta - \hat{\beta})^{\mathrm{T}}F^{\mathrm{T}}F(\beta - \hat{\beta}) \leqslant ps^2 F_{p,v,\alpha} \tag{5.23}$$

where s^2 is an estimate of σ^2 on v degrees of freedom and $F_{p,v,\alpha}$ is the α per cent point of the F distribution on p and v degrees of freedom.

In the p-dimensional space of the parameters, (5.23) defines an ellipsoid, the boundary of which is the contour of constant residual sum of squares

$$S(\beta) - S(\hat{\beta}) = ps^2 F_{p,v,\alpha}.$$

The volume of the ellipsoid is inversely proportional to the square root of the determinant $|F^{\mathrm{T}}F|$. For the single slope parameter in simple regression, the variance, given by (5.7), is minimized if $\Sigma(x_i - \bar{x})^2$ is large. From (5.22), $|(F^{\mathrm{T}}F)^{-1}| = 1/|F^{\mathrm{T}}F|$ is called the generalized variance. Designs which maximize $|F^{\mathrm{T}}F|$ are called D-optimum (for determinant). They are discussed in Chapter 9 and are the subject of Chapter 11.

As well as the volume, the shape of the confidence region depends upon $F^{\mathrm{T}}F$. The implications of various shapes of confidence region, and their dependence on the experimental design, are described in Chapter 6. Several criteria for optimum experimental design and their relationship to confidence regions are discussed in Chapter 10.

The predicted value of the response, given by (5.11) for simple regression, becomes

$$\hat{y}(x) = f^{\mathrm{T}}(x)\hat{\beta} \tag{5.24}$$

when β is a vector, with variance

$$\text{var}\{\hat{y}(x)\} = \sigma^2 f^T(x)(F^T F)^{-1} f(x). \tag{5.25}$$

These formulae are now exemplified for the designs and models of §5.2.

***Example* 5.1** Simple regression (continued)

For simple regression (5.4), the information matrix is

$$F^T F = \begin{bmatrix} \Sigma 1 & \Sigma x_i \\ \Sigma x_i & \Sigma x_i^2 \end{bmatrix} = \begin{bmatrix} N & \Sigma x_i \\ \Sigma x_i & \Sigma x_i^2 \end{bmatrix},$$

where again all summations are over $i = 1, \ldots, N$. The determinant of the information matrix is

$$\begin{aligned} |F^T F| &= \begin{vmatrix} N & \Sigma x_i \\ \Sigma x_i & \Sigma x_i^2 \end{vmatrix} \\ &= N\Sigma x_i^2 - (\Sigma x_i)^2 \\ &= N\Sigma(x_i - \bar{x})^2. \end{aligned} \tag{5.26}$$

Thus the covariance matrix of the least squares estimates $\hat{\beta}_0$ and $\hat{\beta}_1$ is

$$\sigma^2(F^T F)^{-1} = \frac{\sigma^2}{|F^T F|} \begin{bmatrix} \Sigma x_i^2 & -\Sigma x_i \\ -\Sigma x_i & N \end{bmatrix} \tag{5.27}$$

where each element is to be multiplied by $\sigma^2/|F^T F|$. In particular, the variance of $\hat{\beta}_1$, which is the (2, 2) element of (5.27), reduces to (5.7).

For the three-point Design 5.1,

$$F^T F = \begin{bmatrix} 3 & 0 \\ 0 & 2 \end{bmatrix} \qquad |F^T F| = 6$$

$$(F^T F)^{-1} = \begin{bmatrix} 1/3 & 0 \\ 0 & 1/2 \end{bmatrix}. \tag{5.28}$$

For this symmetric design the estimates of the parameters are uncorrelated, whereas for Design 5.2, with only two support points,

$$F^T F = \begin{bmatrix} 3 & 1 \\ 1 & 3 \end{bmatrix} \qquad |F^T F| = 8$$

$$(F^T F)^{-1} = \begin{bmatrix} 3/8 & -1/8 \\ -1/8 & 3/8 \end{bmatrix}. \tag{5.29}$$

Thus the two estimates are negatively correlated.

From (5.25) the variance of the predicted response from Design 5.1 is

$$\frac{\text{var}\{\hat{y}(x)\}}{\sigma^2} = (1 \quad x)\begin{bmatrix} 1/3 & 0 \\ 0 & 1/2 \end{bmatrix}\begin{pmatrix} 1 \\ x \end{pmatrix} = \frac{1}{3} + \frac{x^2}{2}.$$

In comparing designs it is often helpful to scale the variance for σ^2 and the number of trials and to consider the standardized variance

$$d(x, \xi) = N\frac{\text{var}\{\hat{y}(x)\}}{\sigma^2} = 1 + \frac{3x^2}{2}. \tag{5.30}$$

This quadratic has a maximum value over the design region $\mathcal{X}$ of 2.5 at $x = \pm 1$. In contrast, the standardized variance for the non-symmetric Design 5.2 is

$$d(x, \xi) = \tfrac{3}{8}(3 - 2x + 3x^2), \tag{5.31}$$

a non-symmetric function which has a maximum over $\mathcal{X}$ of 3 at $x = -1$.

These numerical results are summarized in Table 5.3. If $|F^{\mathrm{T}}F|$ is to be used to select a design for the first-order model, Design 5.2 is preferable. If, however, the criterion is to minimize the maximum of the standardized variance $d(x, \xi)$ over $\mathcal{X}$, a criterion known as G-optimality, Design 5.1 would be selected. This example shows that a design which is optimum for one purpose may not be so for another. The General Equivalence Theorem of Chapter 9 establishes a relationship between G-optimality and D-optimality. $\qquad\square$

Table 5.3. Determinants and variances for designs for a single quantitative factor

	Number of trials N	$|F^{\mathrm{T}}F|$	max $d(x, \xi)$ $\mathcal{X}$
First-order model			
Design 5.1	3	6	2.5
Design 5.2	3	8	3
Quadratic model			
Design 5.1	3	4	3
Design 5.3	4	7.0233	3.8144

Example **5.2** Quadratic regression (continued)

For the quadratic regression model in one variable (5.16)

$$F^{\mathrm{T}}F = \begin{bmatrix} N & \Sigma x_i & \Sigma x_i^2 \\ \Sigma x_i & \Sigma x_i^2 & \Sigma x_i^3 \\ \Sigma x_i^2 & \Sigma x_i^3 & \Sigma x_i^4 \end{bmatrix}$$

The symmetric three-point Design 5.1 yields

$$F^{T}F = \begin{bmatrix} 3 & 0 & 2 \\ 0 & 2 & 0 \\ 2 & 0 & 2 \end{bmatrix} \qquad (5.32)$$

with $|F^{T}F| = 4$ and

$$(F^{T}F)^{-1} = \begin{bmatrix} 1 & 0 & -1 \\ 0 & 1/2 & 0 \\ -1 & 0 & 3/2 \end{bmatrix}$$

Now $f^{T}(x) = (1 \ x \ x^2)$ and, from (5.32), the standardized variance

$$d(x, \xi) = 3 - 9x^2/2 + 9x^4/2. \qquad (5.33)$$

This symmetric quartic has a maximum over $\mathscr{X}$ of 3 at $x = -1, 0,$ or 1, which are the three design points. Further, this maximum value is equal to the number of parameters p.

Design 5.3 is again symmetric, but $N = 4$ and

$$F^{T}F = \begin{bmatrix} 4 & 0 & 20/9 \\ 0 & 20/9 & 0 \\ 20/9 & 0 & 164/81 \end{bmatrix}$$

with $|F^{T}F| = 7.0233$. The standardized variance is

$$d(x, \xi) = 2.562 - 3.811x^2 + 5.062x^4,$$

again a symmetric quartic, but now the maximum value over $\mathscr{X}$ is 3.814 when $x = \pm 1$.

These results for the second-order model, summarized in Table 5.3, again seem to indicate that the two designs are better for different criteria. However, Design 5.1 is for three trials, and Design 5.3 is for four. The variances $d(x, \xi)$ are scaled to allow for this difference in N. To scale $|F^{T}F|$ for Design 5.3 we multiply by $(3/4)^3$, obtaining 2.963. Thus, on a per trial basis, Design 5.1 is preferable to Design 5.3 for the quadratic model. The implications of comparisons of design per trial are explored in Chapter 9 when we consider exact and approximate designs. □

Example 5.3 Quadratic regression with a single qualitative factor (continued)
The least squares results of this section extend straightforwardly to the model

for quantitative and qualitative factors (5.17). Replication of Design 5.1 at the two levels of z yields the information matrix

$$W^{\mathrm{T}}W = \begin{bmatrix} 3 & 0 & 0 & 2 \\ 0 & 3 & 0 & 2 \\ 0 & 0 & 4 & 0 \\ 2 & 2 & 0 & 4 \end{bmatrix}$$

which is related to the structure of (5.32). In general, the upper $l \times l$ matrix is diagonal for such balanced designs. The 2×2 lower right submatrix results from the two replications of the design for the quantitative factors. This structure is important for designs with both qualitative and quantitative factors described in Chapter 13.   □

5.4 Further reading

Least squares and regression are described in many introductory books on statistics, for example Newbold (1988, Chapters 12–14). A more advanced treatment, whilst remaining firmly rooted in applications, is given by Weisberg (1985). Similar material, at greater length, can be found in Draper and Smith (1981). The more mathematical aspects of the subject are well covered by Seber (1977).

6
Criteria for a good experiment

6.1 Aims of a good experiment

The results of Chapter 5 illustrate that the variances of the estimated parameters in a linear model depend upon the experimental design, as does the variance of the predicted response. An ideal design would provide small values of both variances. However, as the results of Table 5.3 show, a design which is good for one property may be less good for another. Usually one or a few important properties are chosen and designs found which are optimum for these properties. In this chapter we first list some desirable properties of an experimental design. We then illustrate the dependence of the ellipsoidal confidence regions and of the variance of the predicted response on the design. Finally, the criteria of D-, G-, and V-optimality are described, and examples of optimum designs are given for simple regression and quadratic regression.

Box and Draper (1975, 1987, Chapter 14) list 14 aims in the choice of an experimental design. Any, all, or some of these properties of a response surface design may be important.

1. Generate a satisfactory distribution of information throughout the region of interest, which may not coincide with the design region $\mathcal{X}$.

2. Ensure that the fitted value, $\hat{y}(x)$ at x, be as close as possible to the true value $\eta(x)$ at x.

3. Make it possible to detect lack of fit.

4. Allow estimation of transformations of both the response and the quantitative experimental factors.

5. Allow experiments to be performed in blocks.

6. Allow designs of increasing order to be built up sequentially. Often, as in Fig. 3.1, a second-order design will follow one of first order.

7. Provide an internal estimate of error from replication.

8. Be insensitive to wild observations and to violation of the usual normal theory assumptions.

9. Require a minimum number of experimental runs.

10. Provide simple data patterns that allow ready visual appreciation.

11. Ensure simplicity of calculation.

12. Behave well when errors occur in the settings of the experimental variables.
13. Not require an impractically large number of levels of the experimental factors.
14. Provide a check on the 'constancy of variance' assumption.

Different aims will, of course, be of different relative importance as circumstances change. Thus point 11, requiring simplicity of calculation, will not much matter if good software is available for the analysis of the experimental results. But, in this context, 'good' implies the ability to check that the results have been correctly entered into the computer. The restriction on the number of levels of the variables (point 13) is likely to be of particular importance when experiments are carried out by unskilled personnel, for example on a production process.

This list of aims will apply for most experiments. Two further aims which may be important for experiments with quantitative factors are as follows.

15. Orthogonality: the designs have a diagonal information matrix, leading to uncorrelated estimates of the parameters.
16. Rotatability: the variance of $\hat{y}(x)$ depends only on the distance from the centre of the experimental region.

Orthogonality is too restrictive a requirement to be attainable in most of the examples considered in this book. However, it is a property of many commonly used designs such as the 2^m factorials and designs for qualitative factors. Rotatability has been much used by Box and Draper (1963) in the construction of designs for second- and third-order response surface models.

6.2 Confidence regions and the variance of prediction

For the moment, of the many objectives of an experiment, we concentrate on the relationship between the experimental design, the confidence ellipsoid for the parameters given by (5.23), and the variance of the predicted response (5.25). Several of the aims listed in §6.1 will be used in later chapters to assess and compare designs.

Table 6.1 gives eight designs for varying N, the first six of which we compare for simple regression, i.e. for the first-order model in one quantitative factor. Suppose that the fitted model is

$$\hat{y}(x) = 16 + 7.5x. \tag{6.1}$$

Contour plots for the parameter values β for which

Table 6.1. Some designs for first- and second-order models when the number of factors $m = 1$

Design	Number of trials N	Values of x
6.1	3	-1 0 1
6.2	6	-1 -1 0 0 1 1
6.3	8	-1 -1 -1 -1 -1 -1 1 1
6.4	5	-1 -0.5 0 0.5 1
6.5	7	-1 -1 -0.9 -0.85 -0.8 -0.75 1
6.6	2	-1 1
6.7	4	-1 -1 0 1
6.8	4	-1 0 0 1

$$(\beta - \hat{\beta})^{\mathrm{T}} F^{\mathrm{T}} F(\beta - \hat{\beta}) = \delta^2 = 1$$

are given in Figs 6.1–6.6 for the first six designs of the table. These sets of elliptical contours are centred at $\hat{\beta} = (16, 7.5)^{\mathrm{T}}$. Comparison of Figs 6.1 and 6.2 shows how increasing the number of points decreases the size of the confidence region: Design 6.2 is two replicates of Design 6.1. These designs are orthogonal, so that $F^{\mathrm{T}} F$ is diagonal and the axes of the ellipses are parallel to the co-ordinate axes. The designs in Figs 6.3 and 6.5 have several trials at or near the lower end of the design region. As a result the designs are not orthogonal: $F^{\mathrm{T}} F$ has non-zero off-diagonal elements and the axes of the ellipses do not lie along the co-ordinate axes. These axes are also of markedly different lengths. The effect is that some contrasts in the parameters are estimated with small variance, and others are estimated much less precisely. The designs in Figs 6.4 and 6.6 are again orthogonal: the two-trial Design 6.6 yields the largest region of all since it is the design with fewest trials.

For designs when more than two parameters are of importance, the ellipses of Figs 6.1–6.6 are replaced by ellipsoids. Graphical methods of assessment then need to be replaced by analytical methods. In general, the lengths of the axes of the ellipsoid are proportional to the square roots of the eigenvalues of $(F^{\mathrm{T}} F)^{-1}$, which are the reciprocals of the eigenvalues of $F^{\mathrm{T}} F$. A design in which the eigenvalues differ appreciably will typically produce long thin ellipsoids. The determinant of $F^{\mathrm{T}} F$ is equal to the product of the eigenvalues of $F^{\mathrm{T}} F$. Hence, in terms of these eigenvalues, a good design should have a large product, giving a confidence region of small content, with the values all reasonably equal. These ideas are formalized in Chapter 10.

We now consider the standardized variance of the predicted response

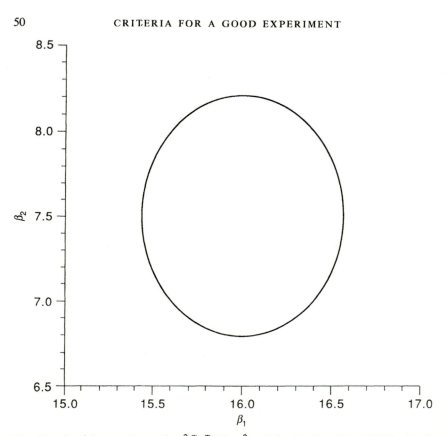

Fig. 6.1. Confidence ellipse $(\beta-\hat{\beta})^{\mathrm{T}}F^{\mathrm{T}}F(\beta-\hat{\beta})=1$ for Design 6.1 of Table 6.1 for simple regression. The ellipse is centred at $\hat{\beta}=(16, 7.5)^{\mathrm{T}}$. This figure arises from a symmetrical design.

$d(x, \xi)$ defined by (5.30). Figures 6.7–6.12 are for the six designs in Table 6.1 which gave rise to the ellipses of Figs 6.1–6.6 The symmetrical designs 6.1, 6.2, 6.4, and 6.6 all give symmetrical plots over $\mathcal{X}$. Because the variances are standardized by the number of trials, Fig. 6.7 is the same as Fig. 6.8. Of these symmetrical designs, Fig. 6.10 shows the highest variance of the predicted response over all $\mathcal{X}$, except for $d(0, \xi)$ which is unity for all four symmetrical designs. The plots in Figs 6.9 and 6.11 illustrate how increasing the number of trials in one area of $\mathcal{X}$, in this case near $x=-1$, reduces the variance in that area but leads to an inflated variance elsewhere. Design 6.4 has its five trials spread uniformly over $\mathcal{X}$, but, as Fig. 6.10 shows, that does not lead to the estimate of $\hat{y}(x)$ with smallest variance over all of $\mathcal{X}$. Figure 6.12 shows that Design 6.6, which has equal numbers of trials at each end of the interval and none at the centre, leads to the design

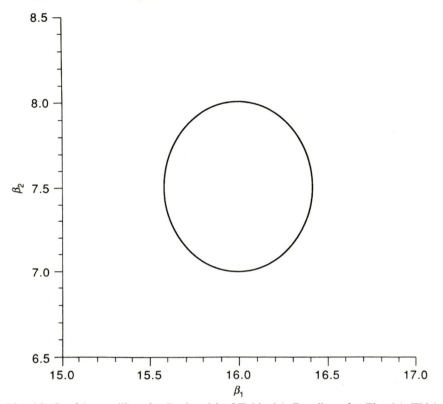

Fig. 6.2. Confidence ellipse for Design 6.2 of Table 6.1. Details as for Fig. 6.1. This figure arises from a symmetrical design.

with smallest $d(x, \xi)$ over the whole of $\mathcal{X}$. We show in §6.3 that no design can do better than this for the first-order model.

When the model is second order in one factor, $d(x, \xi)$ becomes a quartic. Figures 6.13–6.17 give plots of this variance for some of the designs of Table 6.1, including some of those used for the plots for the first-order model. As can be seen in Fig. 6.13, Design 6.1 ensures that the maximum of $d(x, \xi)$ over the design region is equal to 3 which is the number of parameters in the model. Kiefer and Wolfowitz (1960) show that this is the smallest possible value. Design 6.4, the design for Fig. 6.14, distributes five trials uniformly over the design region. The maximum value of $d(x, \xi)$ is greater than that in Fig. 6.13, but the variance is reduced in the centre of $\mathcal{X}$. Locating the trials mainly at one end of the interval causes the variance to be large elsewhere, as in Fig. 6.15 which results from Design 6.5. The remaining two figures are produced by designs in which one trial is added

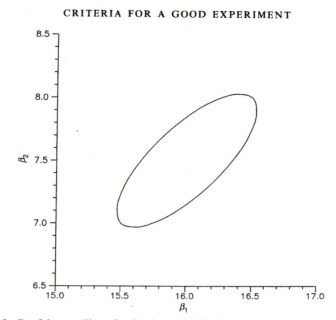

Fig. 6.3. Confidence ellipse for Design 6.3 of Table 6.1. Details as for Fig. 6.1.

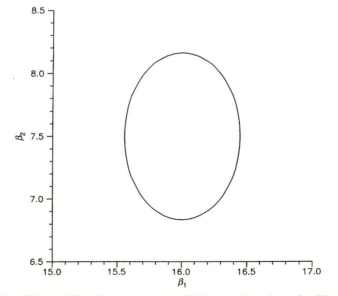

Fig. 6.4. Confidence ellipse for Design 6.4 of Table 6.1. Details as for Fig. 6.1. This figure arises from a symmetrical design.

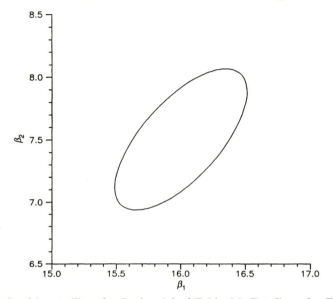

Fig. 6.5. Confidence ellipse for Design 6.5 of Table 6.1. Details as for Fig. 6.1.

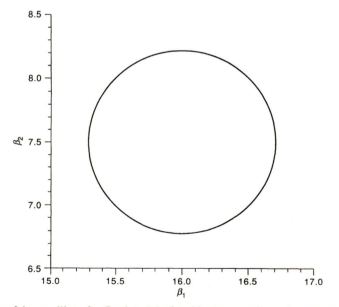

Fig. 6.6. Confidence ellipse for Design 6.6 of Table 6.1. Details as for Fig. 6.1. This figure arises from a symmetrical design.

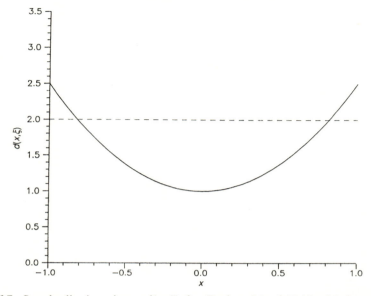

Fig. 6.7. Standardized variance $d(x, \xi)$ for Design 6.1 of Table 6.1 for simple regression.

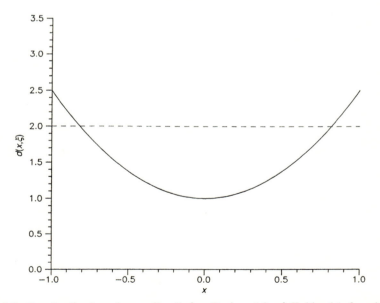

Fig. 6.8. Standardized variance $d(x, \xi)$ for Design 6.2 of Table 6.1 for simple regression. This figure is identical with Fig. 6.7 because Design 6.2 is two replicates of Design 6.1.

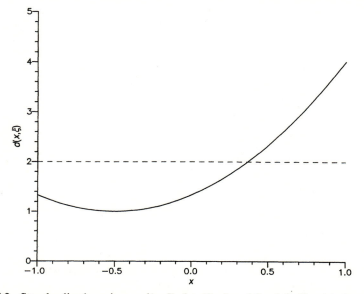

Fig. 6.9. Standardized variance $d(x, \xi)$ for Design 6.3 of Table 6.1 for simple regression. Design 6.3 has several trials at or near -1.

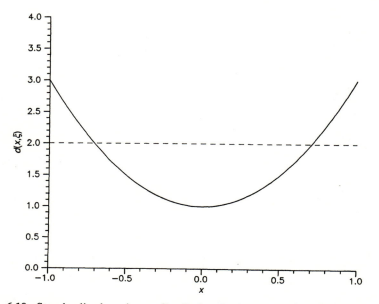

Fig. 6.10. Standardized variance $d(x, \xi)$ for Design 6.4 of Table 6.1 for simple regression.

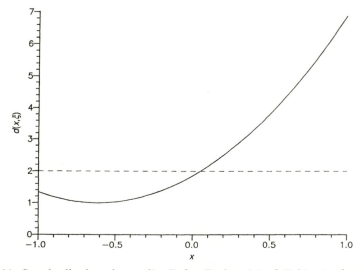

Fig. 6.11. Standardized variance $d(x, \xi)$ for Design 6.5 of Table 6.1 for simple regression. Design 6.5 has several trials at or near -1.

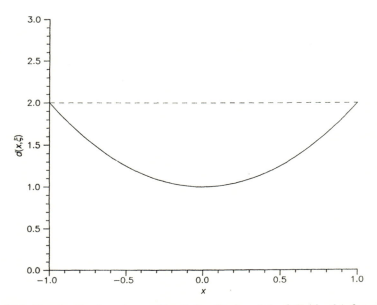

Fig. 6.12. Standardized variance $d(x, \xi)$ for Design 6.6 of Table 6.1 for simple regression. Design 6.6 is D- and G-optimum; the maximum value of $d(x, \xi)$ is 2.

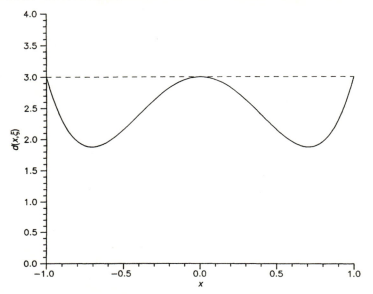

Fig. 6.13. Standardized variance $d(x, \xi)$ for the three-point Design 6.1 and the quadratic model. The design is D- and G-optimum, and the maximum value of $d(x, \xi)$ is 3, the number of parameters in the model.

to the design for Fig. 6.13. In the first case the added trial is at $x = -1$, whereas in the other it is at $x = 0$. The two resulting plots for the variances are quite different. In the first case the variance is reduced in the area near to the lower boundary of $\mathscr{X}$, while it is increased elsewhere. The design for Fig. 6.17 ensures low variance in the centre of the region in which the factor varies, i.e. near the replicated point. However, $d(x, \xi)$ increases sharply towards the ends of the region.

6.3 Some criteria for optimum experimental designs

The examples in this chapter and in Chapter 5 show how different designs can be in the values they yield of $|F^{\mathrm{T}}F|$, in the curve of $d(x, \xi)$ over the design region $\mathscr{X}$, and in the maximum value of the variance over $\mathscr{X}$. An ideal design for these models would simultaneously minimize the generalized variance of the parameter estimates and minimize $d(x, \xi)$ over $\mathscr{X}$. Usually a choice has to be made between these desiderata. Three possible design criteria which relate to these properties are as follows.

D-optimality: a design is D-optimum if it maximizes the value of $|F^{\mathrm{T}}F|$, i.e. the generalized variance of the parameter estimates is minimized.

G-optimality: a G-optimum design minimizes the maximum over the design

CRITERIA FOR A GOOD EXPERIMENT

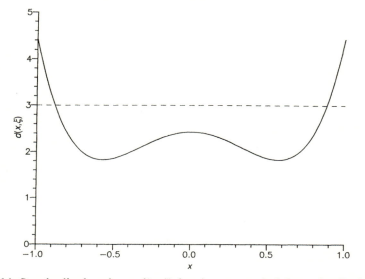

Fig. 6.14. Standardized variance $d(x, \xi)$ for the symmetrical five-point Design 6.4 and quadratic regression.

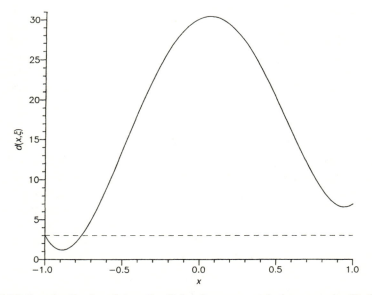

Fig. 6.15. Standardized variance $d(x, \xi)$ for the asymmetrical seven-point Design 6.5 and quadratic regression.

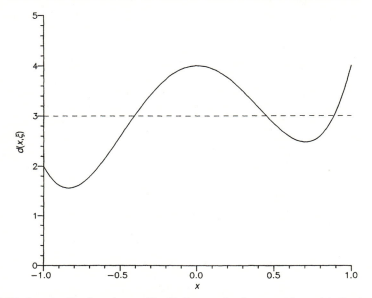

Fig. 6.16. Standardized variance $d(x, \xi)$ for quadratic regression with Design 6.7. Design 6.1 plus one trial at $x = -1$.

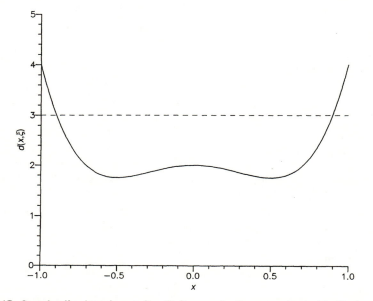

Fig. 6.17. Standardized variance $d(x, \xi)$ for quadratic regression with Design 6.8. Design 6.1 plus one trial at $x = 0$.

region $\mathcal{X}$ of the standardized variance $d(x, \xi)$. This maximum value equals p.

V-optimality: an alternative to *G*-optimality is *V*-optimality in which the average of $d(x, \xi)$ over $\mathcal{X}$ is minimized.

The mathematical construction and evaluation of designs according to the criteria of *D*- and *G*-optimality, and the close relationship between these criteria, are the subjects of Chapter 9. More general criteria, including *V*-optimality, are described in Chapter 10. We conclude this chapter by putting together the numerical results obtained so far for first- and second-order polynomials in one variable.

***Example* 6.1 (*Example* 5.1 *continued*)** Simple regression

The fitted simple regression model was written in (5.11) as

$$\hat{y}(x) = \bar{y} + \hat{\beta}_1 (x - \bar{x}).$$

This suggests rewriting model (5.4) with centred x as

$$E(Y) = \alpha + \beta_1 (x - \bar{x}), \tag{6.2}$$

which again gives (5.11) as a fitted model. With the model written in the orthogonal form (6.2), the diagonal information matrix has the revealing structure

$$F^T F = \begin{pmatrix} N & 0 \\ 0 & \Sigma(x_i - \bar{x})^2 \end{pmatrix}$$

so that the *D*-optimum design is that which minimizes the variance of $\hat{\beta}_1$ (5.7), with the value of N being fixed. If $\mathcal{X} = [-1, 1]$, $\Sigma(x_i - \bar{x})^2$ is maximized by putting half the trials at $x = +1$ and the other half at $x = -1$. When $N = 2$, this is Design 6.6. The plot of $d(x, \xi)$ in Fig. 6.12 has a maximum value over $\mathcal{X}$ of 2, the smallest possible maximum. Therefore this design is also *G*-optimum.

Provided that N is even, replications of this design are *D*- and *G*-optimum. If N is odd, the situation is more complicated. The results of Section 5.3 for $N = 3$, summarized in Table 5.3, show that Design 5.2, in which one of the points $x = 1$ or $x = -1$ is replicated, is to be preferred for *D*-optimality. However, the three-point Design 5.1 is preferable on grounds of *G*-optimality, giving rise to the symmetric curve of $d(x, \xi)$ in Fig. 6.7 with a maximum of 2.5.

A design in which the distribution of trials over $\mathcal{X}$ is specified by a measure ξ, regardless of N, is called continuous or approximate. The equivalence of *D*- and *G*-optimum designs holds, in general, only for continuous designs. Designs for a specified number of trials are called exact.

In this example the exact D-optimum design for even N puts $N/2$ trials at $x = -1$ and $N/2$ at $x = +1$. If N is not even (e.g. 3 or 5), this equal division is, of course, not possible and the exact optimum design will depend on the value of N. Often, as here, a good approximation to the optimum exact design is found by integer approximation to the continuous design. For $N = 5$, two trials could be put at $x = -1$ and three at $x = +1$, or vice versa. But, in more complicated cases, especially when N is barely greater than the number of parameters in the model, the optimum exact design has to be found by numerical optimization for each N. Algorithms for the construction of exact designs are the subject of Chapter 15.   □

Example **6.2** (*Example* **5.2** *continued*) Quadratic regression

For the one-factor quadratic model (5.16), Fig. 6.13 is a plot of $d(x, \xi)$ for the three-point symmetric Design 5.1 or 6.1. The maximum value of $d(x, \xi)$ is 3 and this design is D- and G-optimum. Again, replication of the design will provide the optimum design provided that N is divisible by 3. For $N = 4$, the D- and G-optimum designs are not the same, an example discussed further in Chapter 9.   □

As well as showing the distinction between exact and continuous designs, these examples serve to display some properties of D-optimum designs. One is that the design depends upon the model. A second is that the number of distinct values of x is often equal to the number of parameters p in the model. Therefore the designs provide no method of checking goodness of fit. The introduction of trials at extra x values in order to provide checks reduces the efficiency for the primary purposes of parameter and response estimation. Designs achieving a balance between parameter estimation and model-checking can be found using the composite design criteria described in Chapter 21.

7
Standard designs

7.1 Introduction

The examples in Chapter 6 show that it is not always possible to find a design which is simultaneously optimum for several criteria. When there are only one or two factors, plots of the variance function, like those in Chapter 6, can be used to assess different aspects of a design's behaviour. But when there are several factors and the model is more complicated, such plots cannot be drawn. Designs satisfying one or several criteria are then found by numerical optimization of the design criterion, a function of the design points and of the number of trials. Algorithms for the construction of continuous designs, which do not depend on the total number of trials, are described in Chapter 9. The more complicated algorithms for exact designs are the subject of Chapter 15. In this chapter we give a short survey of some standard designs which avoid the use of these algorithms, particularly when the factors are all quantitative. If particular properties of the design are important, the equivalence theorems of Chapter 10 can be usd to assess and compare designs.

7.2 The 2^m factorial

The design consists of all combinations of points at which the factors take coded values of -1 or 1. The most complicated model that can be fitted to the results contains first-order terms in all factors and two-factor and higher-order interactions up to that of order m. For example, if $m = 3$, the model is

$$E(Y) = \beta_0 + \beta_1 x_1 + \beta_2 x_2 + \beta_3 x_3 + \beta_{12} x_1 x_2 + \beta_{13} x_1 x_3 + \beta_{23} x_2 x_3 + \beta_{123} x_1 x_2 x_3.$$

$$(7.1)$$

The design is given in standard order in Table 7.1. The order of these treatment combinations should be randomized when the design is used. Two notations for the design are used in the table. In the second the factors are represented by capital letters; the presence of the corresponding lower-case letter in the treatment combination indicates that the factor is at its high level. This notation is particularly useful when blocking 2^m factorial designs and for the construction of fractional factorials.

The 2^m designs are easy to construct and easy to analyse. The designs are orthogonal, so that the information matrix $F^T F$ is diagonal. Each diagonal element has the value $N = 2^m$. Because of the orthogonality of the design each

Table 7.1. 2^m factorial design in three factors

Trial number	Factors			Treatment combination	Response
	A x_1	B x_2	C x_3		
1	-1	-1	-1	(1)	$y_{(1)}$
2	$+1$	-1	-1	a	y_a
3	-1	$+1$	-1	b	y_b
4	$+1$	$+1$	-1	ab	y_{ab}
5	-1	-1	$+1$	c	y_c
6	$+1$	-1	$+1$	ac	y_{ac}
7	-1	$+1$	$+1$	bc	y_{bc}
8	$+1$	$+1$	$+1$	abc	y_{abc}

treatment effect can be estimated independently of any other. Further, the estimates have a simple form: since each column of the extended design matrix F consists of $N/2$ elements equal to $+1$ and the same number equal to -1, the elements of $F^T y$ consists of differences of sums of particular halves of the observations. For example, in the notation of Table 7.1,

$$\hat{\beta}_1 = \tfrac{1}{8}\{(y_a + y_{ab} + y_{ac} + y_{abc}) - (y_{(1)} + y_b + y_c + y_{bc})\}. \tag{7.2}$$

Thus the estimate depends on the difference in the average response at the high level of A (treatment combinations including the symbol a) and that at the low level (treatments without the symbol a). This structure extends to estimation of any of the coefficients in the model (7.1). Estimation of $\hat{\beta}_{12}$ requires the vector $x_1 x_2$, found by multiplication from Table 7.1 of the columns for x_1 and x_2. In order, the elements are

$$(+1 \ -1 \ -1 \ +1 \ +1 \ -1 \ -1 \ +1),$$

so that

$$\hat{\beta}_{12} = \tfrac{1}{8}\{(y_{(1)} + y_{ab} + y_c + y_{abc}) - (y_a + y_b + y_{ac} + y_{bc})\}. \tag{7.3}$$

The first group of four units in (7.3) are those with an even parity of treatment letters with ab (either 0 or 2) whereas the second group have an odd parity, in this case one letter in common with ab.

This structure extends straightforwardly to higher values of m. So does the D-optimality of the design. The maximum value of the variance of the predicted response is at the corners of the experimental region, where all x_j^2 have a value of unity. Since the variance of each parameter estimate is σ^2/N, the maximum value of the standardized variance $d(x, \xi)$ is p at each design point.

So far we have assumed that the full model with all interaction terms is to be fitted. Once the model has been fitted, the parameter estimates can be tested for significance. The omission of non-significant terms then leads to simplification of the model. An example of such an analysis, in the absence of an independent estimate of error, is given as Example 8.3 in the next chapter.

A second assumption is that the full factorial model is adequate. However, some curvature may be present in the quantitative factors, which would require the addition of quadratic terms to the model. In order to check whether such terms are needed, three or four 'centre points' are often added to the design. These extra experiments at $x_j = 0$ $(j = 1, \ldots, m)$ also provide a few degrees of freedom for the estimation of σ^2. The concept of a centre point is usually not meaningful for qualitative factors, any more than is the idea of curvature. If qualitative factors are present, one strategy is to include centre points in the quantitative factors at each level of the qualitative ones. Further design aspects of checking for evidence of lack of fit are discussed in §20.5.

Factorial designs with factors at more than two levels are much used, particularly for qualitative factors. If some of the factors are at four levels, the device of 'pseudo-factors' can be used to preserve the convenience for blocking and fractionation of the 2^m series. For example, suppose that a qualitative factor has four levels $T_1, \ldots, T_4$. These can be represented by two pseudo-factors, each at two levels:

Level of qualitative factor	T_1	T_2	T_3	T_4
Pseudo-factor levels	(1)	a	b	ab

In interpreting the analysis of such experiments, it needs to be remembered that AB is not an interaction.

7.3 Blocking 2^m factorial experiments

The family of 2^m factorials can readily be divided into 2^f blocks of size 2^{m-f}, customarily at the sacrifice of information on high-level interactions.

Consider again the 2^3 experiment of the preceding section. The estimate of the three-factor interaction is

$$\hat{\beta}_{123} = \tfrac{1}{8}\{(y_a + y_b + y_c + y_{abc}) - (y_{(1)} + y_{ab} + y_{ac} + y_{bc})\}. \tag{7.4}$$

If the experimental units are divided into two groups, as shown in Table 7.2(a), with treatments a, b, c and abc in one block and the remaining four in the other, then any systematic difference between blocks will be estimated by (7.4). The three-factor interaction is said to be confounded with blocks. Because of the orthogonality of the design, all other effects are estimated without any effect from the blocking.

Table 7.2. Blocking of 2^m factorial experiments

(*a*) 2^3 *in two blocks:* $I=\text{ABC}$

Number of symbols in common with ABC	Block number	Treatment combinations			
Odd	1	a	b	c	abc
Even	2	(1)	ab	ac	bc

(*b*) 2^4 *in two blocks:* $I=\text{ABCD}$

Number of symbols in common with ABC	Block number	Treatment combinations							
Odd	1	a	b	c	abc	d	abd	acd	bcd
Even	2	1	ab	ac	bc	ad	bd	cd	abcd

(*c*) 2^5 *in four blocks:* $I=\text{ABC}=\text{CDE}=\text{ABDE}$

Number of symbols in common with ABC	CDE	Block number	Treatment combinations							
Odd	Odd	1	c	abc	ad	bd	ae	be	cde	abcde
Even	Odd	2	ac	bc	d	abd	e	abe	acde	bcde
Odd	Even	3	a	b	cd	abcd	ce	abce	ade	bde
Even	Even	4	1	ab	acd	bcd	ace	bce	de	abde

It is customary to use high-order interactions for blocking. Table 7.2(b) gives a division of the 16 trials of a 2^4 experiment into two blocks of eight such that the four-factor interaction is confounded with blocks. This design thus has as its defining contrast $I=\text{ABCD}$. The two blocks consist of those treatment combinations which respectively have an odd and an even number of characters in common with ABCD.

As a last example we divide the 32 trials of a 2^5 design into four blocks of eight. This is achieved by dividing the trials according to two defining contrasts, when their product is also a defining contrast. We choose the two four-factor interactions ABC and CDE. The third defining contrast is given by

$$I=\text{ABC}=\text{CDE}=\text{ABC}^2\text{DE}=\text{ABDE}, \tag{7.5}$$

the product of any character with itself being the identity. Table 7.2(c) gives the four blocks, which contain all treatments with a specific combination of an odd and even number of characters in common with ABC and CDE. In practice, the choice of which interactions to include in the defining contrast depends upon which interactions are either negligible or not of interest.

7.4 2^{m-f} fractional factorial designs

A disadvantage of the 2^m factorial designs is that the number of trials increases rapidly with m. As a result, very precise estimates are obtained of all parameters, including high-order interactions. If these interactions are known to be negligible, information on the main effects and lower-order interactions can be obtained more economically by running only a fraction of the complete $N = 2^m$ trials.

A half-fraction of the 2^4 design can be obtained by running one of the two blocks of Table 7.2(b). Each effect of the full 2^4 design will now be aliased with another effect in that they are estimated by the same linear combination of the observations. The defining contrast for the factorial in two blocks was $I = ABCD$. The two 2^{4-1} fractional factorials are generated by the relationships $I = -ABCD$ for the design given by the first block in Table 7.2(b) and $I = ABCD$ for the second. The alias structure is found by multiplication into the generator. In this case $I = ABCD$ gives the alias structure

$$A = BCD \qquad B = ACD \qquad C = ABD \qquad D = ABC$$
$$AB = CD \qquad AC = BD \qquad AD = BC.$$

If the three-factor interactions are negligible, unbiased estimates of the main effects are obtained. However, the two-factor interactions are confounded in pairs. Interpretation of the results of fractional experiments is often helped by the observation that interactions between important factors are more likely than interactions between factors which are not individually significant. If interpretation of the estimated coefficients remains ambiguous, further experiments have to be undertaken. In this example, the other half of the 2^{4-1} design would have to be performed.

Running one of the blocks of the 2^5 design in Table 7.2(c) gives a 2^{5-2} factorial, again with eight trials. For this quarter-replicate, each effect is confounded with three others, given by multiplication into the generators of the design. Multiplication into (7.5) gives the alias structure for the main effects in the fourth fraction of Table 7.2(c) as

$$A = BC = ACDE = BDE$$
$$B = AC = BCDE = ADE$$
$$C = AB = DE = ABCDE$$
$$D = ABCD = CE = ABE$$
$$E = ABCE = CD = ABD.$$

For the 2^{5-2} design consisting of the first block of Table 7.2(c), the alias structure follows from the generators

$$I = -ABC = -CDE = ABDE$$

giving, for example,

$$A = -BC = -ACDE = BDE,$$

which is the same structure as before but with some signs changed.

For this design the shortest word amongst the generators has length 3. The design is then said to be of 'resolution 3'. In a resolution 3 design at least some main effects are aliased with two-factor interactions, but no main effects are aliased with each other. In a resolution 4 design some two-factor interactions are aliased with each other but, at worst, main effects are aliased with three-factor interactions.

An alternative method of generating a 2^{m-f} fractional factorial is to start with a factorial in $m-f$ factors and to add f additional factors. For example, the first block of the 2^{4-1} design of Table 7.2(b), for which $D = -ABC$, could be generated by imposing this relationship on a 2^3 factorial. In the alternative notation, which is more convenient for most of this book, this corresponds to putting $x_4 = -x_1 x_2 x_3$. The levels of x_4 are then determined by the level of the three-factor interaction between the factors of the original 2^3 experiment. The second 2^{4-1} fractional design is found by putting $x_4 = x_1 x_2 x_3$.

A design for $m = 6$ and $f = 3$ is shown in Table 7.3. This has generators $x_4 = x_1 x_2 x_3$, $x_5 = x_1 x_2$, and $x_6 = x_2 x_3$. The alias structure is found by multiplying these generators together to give the full set of eight, perhaps more clearly written in letters as

$$I = ABCD = ABE = BCF = CDE = ADF = ACEF = BDEF.$$

Table 7.3. 2^{m-f} factorial design in six factors and $f = 3$: 'first-order design'

N	x_1	x_2	x_3	x_4 $(=x_1 x_2 x_3)$	x_5 $(=x_1 x_2)$	x_6 $(=x_2 x_3)$
1	-1	-1	-1	-1	1	1
2	1	-1	-1	1	-1	1
3	-1	1	-1	1	-1	-1
4	1	1	-1	-1	1	-1
5	-1	-1	1	1	1	-1
6	1	-1	1	-1	-1	-1
7	-1	1	1	-1	-1	1
8	1	1	1	1	1	1

Then, for example, the alias structure for A is

$$A = BCD = BE = ABCF = ACDE = DF = CEF = ABDEF.$$

The design is of resolution 3. In the absence of any two-factor interactions all six main effects may be estimated. Such designs, called main-effect plans or designs, are often used in the screening stage of an experimental programme mentioned in §3.2.

7.5 Composite designs

If a second-order polynomial in m factors is to be fitted, observations have to be taken at more than two levels of each factor, as in the design of Table 3.2. One possibility is to extend the designs of §7.2 to the 3^m factorials which consist of all combinations of each factor at levels -1, 0, and 1. As m increases, such designs rapidly require an excessive number of trials. The composite designs provide a family of efficient designs requiring appreciably fewer trials.

Composite designs consist of the points of a 2^{m-f} fractional factorial, for $f \geqslant 0$, and $2m$ star points. These star points have $m-1$ zero co-ordinates and one equal to either α or $-\alpha$. When the design region is cubic (taken to include both the square and the hypercube) $\alpha = 1$. When the design region is spherical, $\alpha = m^{1/2}$. If $m = 1$, a centre point must be included. However, three or four centre points are often incorporated in the design, whatever the value of m, giving 'central composite designs'. These centre points provide, as they do for 2^m factorials, an estimate of the error variance σ^2 based solely on replication. They also provide a test for lack of fit. If there is evidence of lack of fit, one possibility is to consider fitting a third-order polynomial to the results of further observations. However, it is usually much better to investigate transformation of the response, which often leads to readily interpretable models with a reduced number of parameters. An example is given in Chapter 8.

Central composite designs are widely used in the exploration of response surfaces around the neighbourhood of maxima and minima. The exact value of α and the number of centre points depend upon the design criterion. Box and Draper (1987, p. 512) give a table of values of those design characteristics which yield rotatable central composite designs. The resulting values of α are close to or equal to $m^{1/2}$ over the range $2 \leqslant m \leqslant 8$. The number of centre points in the absence of blocking is in the range 2–4. The effect of centre points is to decrease the efficiency of the designs as measured by D-optimality. A further distinction between the designs produced by the two criteria is that the Box–Draper designs are shrunk away from the edges of the experimental region in order to reduce the effect of bias from higher-order terms omitted from the model. This protection is bought at the cost of reduced efficiency as measured by D- or G-optimality: D-optimum designs span the experimental region, at least for linear models.

To conclude this section we give, in Table 7.4, an example of a five-factor central composite design for a cubic region. The design includes the points of a 2^{5-1} factorial, $2m$ star points at the centres of the faces of the cube formed by the factorial points, and four centre points. The total number of trials is $N = 2^{5-1} + 2 \times 5 + 4 = 30$. This number is great enough to allow estimation of the 21 parameters of the second-order model

$$E(Y) = \beta_0 + \sum_{i=1}^{5} \beta_i x_i + \sum_{i=1}^{4} \sum_{j=i+1}^{5} \beta_{ij} x_i x_j + \sum_{i=1}^{5} \beta_{ii} x_i^2.$$

The generator for the fractional factorial part of the design was taken as $x_5 = x_1 x_2 x_3 x_4$. In practice, the 30 trials of Table 7.4 should be run in random order.

7.6 Catalogues of optimum exact designs

The comparisons of Lucas (1976) and of Donev and Atkinson (1988) show that central composite designs behave well according to the criteria of D- and G-optimality and also for the average variance criterion, V-optimality. However, the designs are restricted to a few values of N. If a design of a rather different size is required, it will either have to be generated by computer search or looked up in a catalogue.

There are several catalogues of exact designs for specified models, usually first or second order. The designs depend on the number of trials N, the number of factors, the design region, and the permitted number of factor levels. As a consequence, the catalogues can be large (e.g. Nalimov 1982). An example of a smaller, more specialized, catalogue is that of Donev (1988) who lists small D-optimum exact designs for second-order polynomials when $m = 2$, 3, 4, and 5, with only three levels of each factor. These designs were found by searching over the points of the 3^m factorial, but require appreciably fewer trials than the full factorial, especially as m increases. Some of these designs are listed in Chapter 11. Donev also gives designs modified by use of the adjustment algorithm of Chapter 15, which searches away from the 3^m grid. The result is an improvement, sometimes small, in D-optimality at the cost of factor settings which can be anywhere in the continuous design region.

A catalogue of sequentially generated designs is given by Vuchkov et al. (1978). Given a design for $N - 1$ trials, the Nth trial is added to maximize the criterion of D-optimality. Although the sequence starts with the D-optimum design for the smallest value of N, the imposed sequential structure may not yield the best design for larger values.

Exact designs with good properties are described by, amongst others, Mitchell and Bayne (1978) and Welch (1984). Some good exact designs for

Table 7.4. Central composite design: $m = 5, f = 1$, cubic region, four centre points

Trial number	x_1	x_2	x_3	x_4	x_5
1	−1	−1	−1	−1	−1
2	1	−1	−1	−1	1
3	−1	1	−1	−1	1
4	1	1	−1	−1	−1
5	−1	−1	1	−1	1
6	1	−1	1	−1	−1
7	−1	1	1	−1	−1
8	1	1	1	−1	1
9	−1	−1	−1	1	1
10	1	−1	−1	1	−1
11	−1	1	−1	1	−1
12	1	1	−1	1	1
13	−1	−1	1	1	−1
14	1	−1	1	1	1
15	−1	1	1	1	1
16	1	1	1	1	−1
17	−1	0	0	0	0
18	1	0	0	0	0
19	0	−1	0	0	0
20	0	1	0	0	0
21	0	0	−1	0	0
22	0	0	1	0	0
23	0	0	0	−1	0
24	0	0	0	1	0
25	0	0	0	0	−1
26	0	0	0	0	1
27	0	0	0	0	0
28	0	0	0	0	0
29	0	0	0	0	0
30	0	0	0	0	0

second-order polynomials are tabulated in Chapter 11. The division of these designs into blocks and the introduction of qualitative factors into response surface designs are considered in Chapters 12–14.

7.7 Further reading

There are several thorough and lucid treatments of 2^m factorials, including blocking and fractional replication. Box *et al.* (1978, Chapters 10–13) use the numerical notation, for example $I = 1234$, which goes more naturally with fitting polynomial models such as (7.1). Davies (1956, Chapters 7, 9, and 10) gives numerous examples of design and analysis with discussion of the physical meaning of the statistical analyses. He uses the alphabetical notation $I = ABCD$ which we find more convenient for generating and describing designs such as those of Table 7.2. A table of blocking arrangements for 2^m factorials is given by Box et al. (1978, pp. 346–7) who also give (p. 410) a table of two-level fractional factorial designs for $m \leqslant 11$ and $N \leqslant 128$. The interpretation of second-order polynomial models is described by Box et al. (1978, Chapter 15) and given in great detail by Box and Draper (1987, Chapters 9–11).

8
The analysis of experiments

8.1 The desorption of carbon monoxide

This book is primarily concerned with the design of experiments rather than with their analysis. However, in this chapter we give analyses of three sets of data to illustrate some points in least squares and the related design criteria.

The second example is an analysis of Derringer's elastomer data from Chapter 1 (Example 1.2) in order to illustrate the importance of transformations of the response in providing simple models. The third example is an analysis of a 2^{5-1} experiment, a class of designs for which there is no obvious estimate of error. Half-normal plots of effects combined with the analysis of variance provide a parsimonious model, i.e. one with few parameters. But we begin with the analysis of a straightforward example with only one explanatory variable, the data on carbon monoxide production given in Table 1.1.

The scatter plot of y against x in Fig. 1.1 seems to show a clear linear relationship between the amount of carbon monoxide desorbed and the initial K/C ratio. The simple linear regression model (5.4) is thus a sensible starting point for modelling these data. Since there are 22 observations at only six distinct values of x, there is appreciable replication, giving a pure error estimate of σ^2 on $22-6=16$ degrees of freedom. Fitting the linear model absorbs two degrees of freedom, so that four degrees of freedom are available for testing the goodness of fit of the regression model. The resulting analysis of variance is given in Table 8.1.

Table 8.1. The desorption of carbon monoxide: analysis of variance for simple regression

Source	Sum of squares	Degrees of freedom	Mean square	F
Regression	42.692	1	42.692	607.7
Lack of fit	0.100	4	0.025	0.36
Pure error	1.124	16	0.070	
Total (corrected)	43.916	21		

The results in the table show no evidence of systematic lack of fit of the model: the observed value of the variance ratio is 0.36, which is less than unity and so cannot be significant. Even if all the remaining structure in the means of the six groups of observations were concentrated into one degree of freedom, the observed value would be 1.42 to be compared with $F_{4,16}$ for which the 5 per cent point is 3.01. The other entries in Table 8.1 show, as expected, a highly significant linear relationship between y and x.

Since there is no evidence of lack of fit, we proceed by pooling, i.e. combining, the four degrees of freedom with the 16 for pure error to give 20 degrees of freedom for error and thus the more standard presentation of the analysis of variance of Table 8.2. In this example pooling these two sources of error has little effect on the analysis. However, more general forms of pooling are important in the analysis of designs with few or no degrees of freedom for pure error, such as the 2^{5-1} design analysed in §8.3. An advantage of pooling in the present example is that standard regression packages use the residual mean square error estimate of σ^2 in the calculation of t tests and standardized residuals. We continue with such a standard analysis.

Table 8.2. The desorption of carbon monoxide: analysis of variance for simple regression with pooled estimate of error

Source	Sum of squares	Degrees of freedom	Mean square	F
Regression	42.692	1	42.692	697.61
Residual (error)	1.224	20	0.061	
Total (corrected)	43.916	21		

Table 8.3 gives the estimated regression coefficient, together with the standard error of the estimate and the t statistic for the hypothesis of no regression (5.9). The value of 26.41 provides overwhelming evidence of a relationship between carbon monoxide production and the initial K/C ratio. The square of this value, 697.6, is the F value for regression in Table 8.2, an identity mentioned in §5.1.

The goodness of fit test of the analysis of variance in Table 8.1 failed to indicate any systematic departure of the means of the groups of observations from linear regression. To check for constancy of variance and for individual outliers we look at plots of residuals and related quantities.

Figure 8.1 is a plot of the least squares residuals e_i (5.13) against the fitted values $\hat{y}_i$. There seems no evidence of a trend in variance, although there are three moderately large residuals belonging to observations 8, 12, and 18.

Table 8.3. The desorption of carbon monoxide: parameter estimates and residuals

---

(a) *Parameter estimates*

Term	Coefficient	Estimated standard error	t-ratio
Constant	−0.0380	0.1004	−0.38
Regression	1.6031	0.0607	26.41

(b) *Residuals*

Observation number i	Residual e_i	Studentized residuals t_i
8	−0.546	−2.26
12	0.484	2.00
18	0.601	2.53

Since least squares residuals do not all have the same variance, comparison between them might be misleading. Although extreme differences in variances are rare, residuals are often standardized to have the same variances. If we rewrite (5.12) as

$$\text{var}\{\hat{y}(x_i)\} = \sigma^2 h_i,$$

then

$$\text{var}(e_i) = \sigma^2 (1 - h_i) \qquad (8.1)$$

and the studentized residuals

$$t_i = \frac{e_i}{\{s^2(1 - h_i)\}^{1/2}} \qquad (8.2)$$

do all have the same variance. Figure 8.2 is a normal plot of studentized residuals, which should be approximately straight if the errors are normally distributed. A problem with interpretation of normal plots is to decide whether a particular plot is sufficiently unstraight to indicate departures from the model. If this is important, simulations of normal data can be used to build up an envelope providing bounds within which the plot should lie. In the present example it is not crucial to establish whether observations 8, 12, and 18

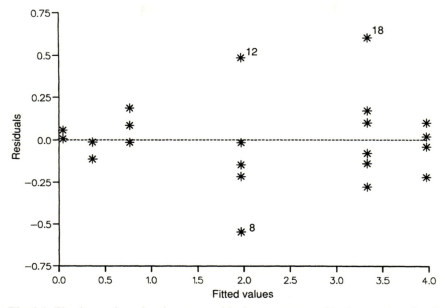

Fig. 8.1. The desorption of carbon monoxide: least squares residuals e_i against fitted values $\hat{y}_i$.

are outliers. The overall relationship between response and explanatory variable is unambiguously established. The major effect of deletion of these three observations is to reduce the estimate of σ^2 and so give slightly smaller confidence intervals for the parameter estimates. However, for these data, too much attention should not be given to the fine structure of the residuals: the numerical values of Table 1.1 were extracted manually from a plot in the original paper and so are subject to non-random patterns in the last digit.

The fitting and checking of this model ends the standard analysis. However, there are some special features of the chemical system which make further analysis interesting.

It might be expected that carbon monoxide will not be desorbed when the K/C ratio is zero, so that the intercept β_0 in (5.4) should be zero. The t test of Table 8.3 does indeed indicate that this is so. The model then becomes

$$\eta(x_i, \beta) = \beta_1 x_i, \qquad (8.3)$$

regression through the origin. However, there is a further simplification. The postulated mechanism for the reaction suggests that $\beta_1 = 1.5$. For the fitted model (8.3), $\hat{\beta}_1 = 1.584$ with an estimated standard error of 0.0312. The t ratio for the hypothesis that $\beta_1 = 1.5$ is therefore 2.67, significant at the 5 per cent level. There is thus some evidence that the true value is not 1.5. A summary of

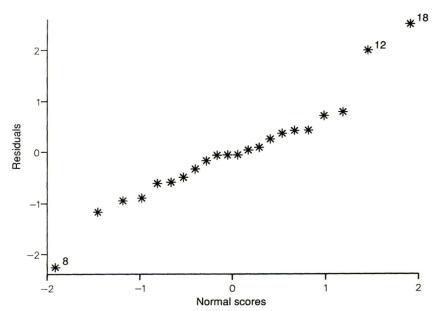

Fig. 8.2. The desorption of carbon monoxide: normal plot of studentized residuals.

these results is given by the analysis of variance of Table 8.4. The F test due to taking $\beta_1 \neq 1.5$ is, to the rounding error, the square of the t statistic.

A final analysis of the data would involve repeating the residual plots of Figs 8.1 and 8.2 for this reduced model. No new points arise. But there are a few general comments on the design of this experiment. If the purpose is solely to estimate the slope of the model with no intercept, the design minimizing the variance of $\hat{\beta}_1$ puts all trials on the boundary of $\mathscr{X}$, i.e. at the maximum value of x. If it is required to check the straight line model against the quadratic alternative

$$\eta(x, \beta) = \beta_1 x + \beta_2 x^2, \tag{8.4}$$

trials at two values of x are required. This design is discussed in Chapter 9.

8.2 Viscosity of elastomer blends

We now consider the analysis of a slightly more complicated data set, the data on the viscosity of elastomer blends from Table 1.2. This gives the viscosity of elastomer–filler blends, for three kinds of filler, as a function of the two quantitative factors, the amount of naphthenic oil, and the amount of filler. For illustration, the analysis concentrates on the 23 results for filler B, which

Table 8.4. The desorption of carbon monoxide: analysis of variance for regression through the origin, testing $\beta_1 = 1.5$

Source	Sum of squares	Degrees of freedom	Mean square	F
$\beta_1 = 1.5$	150.417	—		
Regression	0.420	1	0.420	6.88
Constant	0.009	1	0.009	0.14
Residual	1.224	20	0.061	
Total	152.070	22		

come from a 4×6 factorial with one missing observation. The interesting feature of the analysis is that a more powerful analysis of the data is obtained by working with the logged response rather than with the original viscosities given in Table 1.2.

To begin, we plot the data. Figure 8.3 shows that there is a clear relationship between viscosity and filler level which is given in scaled units. The relationship appears to be slightly curved, with the variance increasing with increasing filler levels. There may also be some interaction between the two quantitative factors. The first-order model

$$\eta(x) = \beta_0 + \beta_1 x_1 + \beta_2 x_2, \tag{8.5}$$

where x_1 is the filler level and x_2 is the level of naphthenic oil, is not satisfactory. The plot of least squares residuals against fitted values for this model (Fig. 8.4) shows clear evidence of the need for curvature terms in the models. The full second-order model

$$\eta(x) = \beta_0 + \beta_1 x_1 + \beta_2 x_2 + \beta_{11} x_1^2 + \beta_{22} x_2^2 + \beta_{12} x_1 x_2 \tag{8.6}$$

yields a residual sum of squares of 201, as opposed to 3821 for the first-order model. The significance of this reduction is calculated in the analysis of variance of Table 8.5, giving a highly significant F value of 102. The inclusion of the second-order terms is justified and, as the t tests of Table 8.6 show, all three extra terms are significant.

Although the second-order model fits the data much better than the first-order model, it is unlikely that observations such as these will be adequately explained by additive errors of constant variance, which are assumed in the method of least squares. The observations are necessarily non-negative and, in this case, have a ratio of just over $10:1$ between the smallest and largest values. Such data are often analysed after transformation to produce homogeneity of variance: for viscosity measurements the logarithmic transformation is frequently used. If the second-order model (8.6) is fitted with log y as the

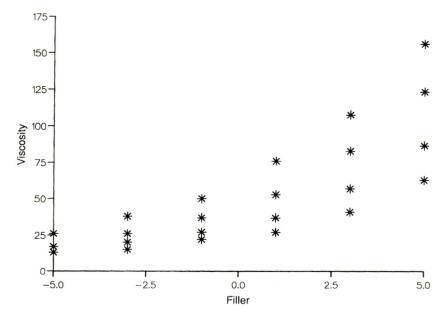

Fig. 8.3. Viscosity of elastomer blends: viscosity against amount of filler (scaled units) for filler **B**.

response, it is found that the second-order terms are not needed. Table 8.6 gives t values for all coefficients of the second-order model for both y and $\log y$ as a response. The difference is amazing. By taking logarithms a simple additive model has been obtained which, in addition, satisfies the constraint that the response must be non-negative. Further evidence for the desirability of the transformation comes from plots of residuals against fitted values, similar to Fig. 8.4, which are not shown here. That for the logged response is structure-free, confirming that all information about the data is in the fitted linear model and the random scatter about that model.

Use of $\log y$ as a response ends the straightforward analysis of these data. However, it is possible that some other transformation might be preferable, for example the reciprocal or the square root of y. It is not possible to assess these alternatives by directly comparing the residual sums of squares, as transformation of the response alters the value, regardless of how well the model fits. For example, the residual sum of squares for $\log y$ for the second-order model is 0.044, to be compared with the value of 201 for the untransformed response in Table 8.5. Therefore we consider the parametric family of power transformations introduced by Box and Cox (1964).

Let $\dot{y}$ be the geometric mean of the observations. Then the response

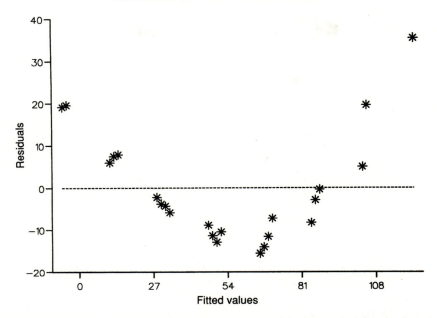

Fig. 8.4. Viscosity of elastomer blends: least squares residuals e_i against fitted values $\hat{y}_i$ for the first-order model (8.5).

Table 8.5. Derringer's elastomer data: analysis of variance

Source	Sum of squares	Degrees of freedom	Mean square	F
First-order terms	27 843	2	13 922	1180
Second-order terms	3620	3	1207	102
Residual	201	17	11.8	
Total (corrected)	31 664	22		

$$z(\lambda) = \begin{cases} \dfrac{y^\lambda - 1}{\lambda \dot{y}^{\lambda-1}} & (\lambda \neq 0) \\ \dot{y} \log y & (\lambda = 0) \end{cases} \tag{8.7}$$

has dimension y for all λ, so that the residual sums of squares of $z(\lambda)$ can be compared directly. For $\lambda = 1$, $z(1)$ is the untransformed response, with $\lambda = 1/2$ the square root of y, $\lambda = 0$ the logarithm, and $\lambda = -1$ the reciprocal. These are probably the values which occur most frequently in the analysis of data.

Table 8.6. Derringer's elastomer data: t values for coefficients of full second-order model

Term	Response	
	y	$\log y$
Constant	25.52	171.05
x_1	38.26	55.98
x_2	−22.05	−31.42
x_1^2	12.64	1.43
x_2^2	3.78	1.04
$x_1 x_2$	−14.07	−0.72

There are several ways of finding a good λ for these data. Figure 8.5 is a plot of the residual sums of squares for the first-order model (continuous line) and for the second-order model (broken line). Clearly, the curve for the second-order model must always be below that for the first-order model. The minimum residual sum of squares for the first-order model is 83.17 at $\lambda = -0.05$, compared with 89.25 at $\lambda = 0$. The F test for the hypothesis of the logarithmic transformation, i.e. $\lambda = 0$, found by comparing sums of squares has the value $(89.25 - 83.17)/(83.17/16) = 1.17$, so that the logarithmic transformation is accepted.

Probably the most interesting feature is the different shape of the two curves. The first-order model is only acceptable over a narrow range of λ values. But the very flat curve for the second-order residual sum of squares means that over a wide range of λ values the addition of second-order terms is an alternative to the logarithmic transformation and a simpler model. This trade-off between transformation and augmenting a linear model is a frequently occurring phenomenon. Often, as here, use of the transformation yields a model with fewer estimated parameters. In this example, a simpler approach to finding a good transformation is to find values of λ for which the first-order model is adequate. Figure 8.6 is a plot of the F statistic for the second-order terms as a function of λ. The first-order model is adequate near zero, but for other values, as we have seen for $\lambda = 1$, there is a highly significant value indicating the need for quadratic and interaction terms. The F value reaches a maximum of over 100 when λ is around 0.9. It then decreases with increasing λ as the residual sum of squares for the second-order model begins to increase, yielding an inflated estimate of the error variance. For larger values of λ than those shown in the plot the F statistic becomes small because neither first-order nor second-order models provide an adequate approxima-

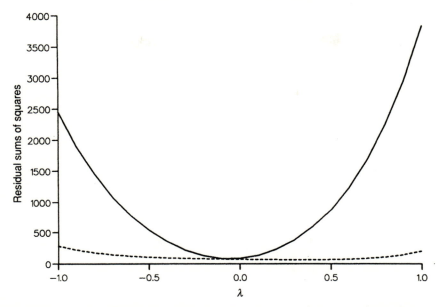

Fig. 8.5. Viscosity of elastomer blends: residuals sums of squares of transformed observations $z(\lambda)$:(——— first-order model (8.5); — — — — second-order model (8.6)).

tion to the data. The simplicity of this approach is that, since the F statistic is a ratio of sums of squares, the units of the transformed response do not matter. Thus, for $\lambda = 0$, the response could be taken as either $\dot{y} \log y$ or $\log y$, both giving the same F value. In general, it is not necessary to calculate $z(\lambda)$ (8.7), but only the relevant powers of y.

The analysis so far has been based on aggregate properties, such as parameter estimates or residual sums of squares, calculated over all the data. However, there remains the possibility that inference about λ is being unduly influenced by one or a few observations. We conclude our analysis of these data by using added variable plots to investigate this possibility.

For a fixed value of λ, which we call λ_0, the need for a transformation can be assessed by Taylor series expansion of $z(\lambda)$ (8.7) to give a 'constructed variable' $w(\lambda_0)$ which is included as an extra explanatory variable in the model. Significant regression on $w(\lambda_0)$ is evidence that a transformation is needed rather than use of λ_0. Often $\lambda_0 = 1$ and the method provides a computationally convenient procedure for assessing whether any transformation is required, which avoids calculation of curves like those of Fig. 8.5. Of course, if there is evidence that a transformation is required, a good value of λ will have to be found, perhaps by numerical search to identify the minimum of curves of

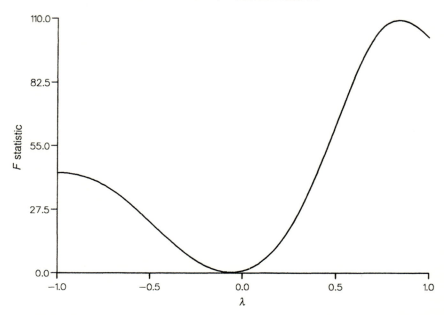

Fig. 8.6. Viscosity of elastomer blends: F test for adequacy of first-order model as the transformation parameter λ varies.

residual sums of squares like those of Fig. 8.5. However, the coefficient of regression on $w(\lambda_0)$ has approximately the value $\lambda_0 - \hat{\lambda}$, where $\hat{\lambda}$ is the value of λ minimizing the residual sum of squares of $z(\lambda)$. Therefore regression on the constructed variable also indicates a suitable transformation.

To obtain information on the contribution of individual observations we use the interpretation of multiple regression as a series of univariate regressions on residuals. In the present case the residuals of the response after regression on the columns of F are

$$e_y = y - F\hat{\beta} = (I - H)y \qquad (8.8)$$

where H is the 'hat' matrix $F(F^\mathrm{T}F)^{-1}F^\mathrm{T}$. Likewise the residuals of the constructed variable after regression on the same F are

$$e_w = (I - H)w.$$

Then the coefficient of regression through the origin of e_y on e_w is identically the coefficient of regression on w in the model also including F. The scatter plot of e_y against e_w is sometimes called a constructed variable plot. For balanced data, such as those of the present example, the plot indicates whether evidence for a transformation is spread throughout the data or whether it is being caused or suppressed by one or a few observations.

For $\lambda_0 = 1$ the constructed variable is

$$w(1) = y\{\log(y/\hat{y}) - 1\}.$$

Figure 8.7 is a plot of residual y against residual $w(1)$. The strong relationship between the two residuals is evidence that a transformation of y is needed, and this evidence could, if desired, be formalized by a t test. All observations obey the same linear relationship; there is no evidence of any important contribution from individual observations.

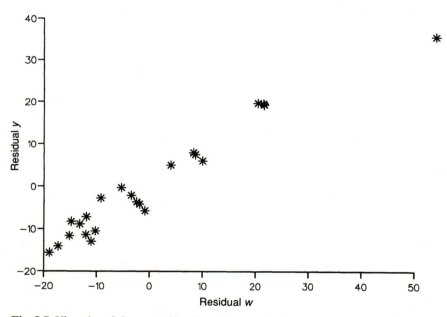

Fig. 8.7. Viscosity of elastomer blends: constructed variable plot for the hypothesis of no transformation ($\lambda_0 = 1$).

We have already seen that the logarithmic transformation is indicated. Figure 8.8 is a plot of residual $z(0)$ against the residual of the constructed variable

$$w(0) = \hat{y} \log y\{(\log y)/2 - \log \hat{y}\}.$$

This constructed variable plot shows a seemingly random scatter of points, with no suggestion of regression, confirming the desirability of the log transformation.

THE ANALYSIS OF EXPERIMENTS

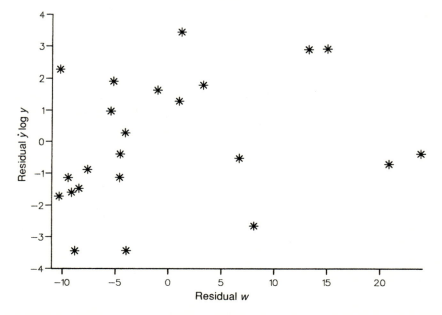

Fig. 8.8. Viscosity of elastomer blends: constructed variable plot for the hypothesis of the log transformation ($\lambda_0 = 0$).

This analysis concludes our discussion of the data of Table 1.2, the greater part of which has been concerned with illustrating the importance of transformations of the response in the analysis of data. We have only looked at part of the example, that to do with the second filler. A more complete analysis of the data of Table 1.2 would investigate transformations of the results for the other two fillers. Although these analyses could proceed separately, a combined analysis yielding one transformation for all three sets of readings would be preferable. Any inferences, such as confidence intervals, should be made on the transformed data, although they may then be back-transformed to the original scale for ease of interpretation.

8.3 Reactor data

This section illustrates the analysis of a 2^{5-1} fractional factorial. In the analysis of the data on the production of carbon monoxide in §8.1 replicate observations were used to provide an estimate of the error variance σ^2. In the analysis of Derringer's elastomer data, the residual sum of squares was used for the same purpose. However, in the analysis of two-level factorials and their fractions, all degrees of freedom can be used to estimate main effects and interactions. Unless there is an external estimate of error, progress is only possible if certain effects can be assumed to be negligible. When this is so the related sums of squares can be pooled and used to provide an estimate of error. In this section we illustrate the use of normal and half-normal lots of effects to indicate which degrees of freedom can be used to estimate error. This indication can then be checked by more formal methods, for example the analysis of variance.

The example is taken from Box *et al.* (1978, p. 377) and is used by them to introduce the idea of a fractional factorial. Data from a full 2^5 factorial are available. However, to make the point that little is lost by using only a half-replicate of the full factorial, Box *et al.* analyse the half the data comprising the 2^{5-1} factorial given in Table 8.7.

Table 8.7. Reactor data: a 2^{5-1} fractional factorial

Factor					-1	$+1$
x_1 feed rate (l/min)					10	15
x_2 catalyst (%)					1	2
x_3 agitation rate (rev/min)					100	120
x_4 temperature (°C)					140	180
x_5 concentration (%)					3	6

x_1	x_2	x_3	x_4	x_5	Response (% reacted) y
−	−	−	−	+	56
+	−	−	−	−	53
−	+	−	−	−	63
+	+	−	−	+	65
−	−	+	−	−	53
+	−	+	−	+	55
−	+	+	−	+	67
+	+	+	−	−	61
−	−	−	+	−	69
+	−	−	+	+	45
−	+	−	+	+	78
+	+	−	+	−	93
−	−	+	+	+	49
+	−	+	+	−	60
−	+	+	+	−	95
+	+	+	+	+	82

The formulae for estimating the parameters of linear models fitted to 2^m factorials and their fractions are given in §7.2. For sufficiently large full factorials it is often assumed that third- and higher-order interactions are negligible. The sums of squares corresponding to these degrees of freedom are then pooled to provide an estimate of error. However, with the 2^{5-1} design all 15 degrees of freedom of this resolution 5 design can be used to estimate either main effects, such as β_1, or two-factor interactions, such as β_{45}. The 16th degree of freedom estimates the mean. In order to determine which degrees of freedom are estimating error, a normal or half-normal plot of the effects, or parameter estimates, is used.

Each parameter estimate is a linear combination, in this example, of 16 observations. If all parameters of the model, apart from the constant term β_0, are zero, each of the 15 estimated coefficients will be a sample from the same distribution. If the observations are normally distributed, the parameter estimates will be also. If the observations are approximately normally distributed, the central limit theorem ensures that the estimates will be more normally distributed than are the individual observations. So, in the null case of no effects, a normal plot of the estimated effects should be a straight line, apart from sampling fluctuations. Any appreciably non-zero parameter will give rise to an estimate which does not follow this distribution and which will be revealed by the plot as an outlier. The remaining estimates which follow the null distribution can then be used to provide the estimate of residual variance. The analysis of variance is often used to confirm this graphically based impression.

More specifically we can write any parameter estimate as

$$\hat{\beta}_j = \sum_{i=1}^{N} \frac{y_i \delta_{ij}}{N} \tag{8.9}$$

where $\delta_{ij} = \pm 1$. Particular examples are (7.2) and (7.3). An equivalent to (8.9) is the 'effect' $2\hat{\beta}_j$, of x_j, which is the change in the estimated response in going from the low to the high level of x_j. Whichever parameterization is used, the sum of squares associated with x_j is

$$SS_j = N\hat{\beta}_j^2, \tag{8.10}$$

the reduction in the residual sum of squares on including x_j in the model. The summation of terms (8.10) for negligible effects provides the residual sum of squares.

Figure 8.9 is a normal plot of the parameter estimates for the data of Table 8.7. The plot shows a relatively straight section, corresponding to effects which can be assumed negligible. There are also five appreciable parameters. In order of magnitude these are, for factors and interactions, 2, 4, 24, 45, and 5. Thus three factors seem to be important, together with two of their

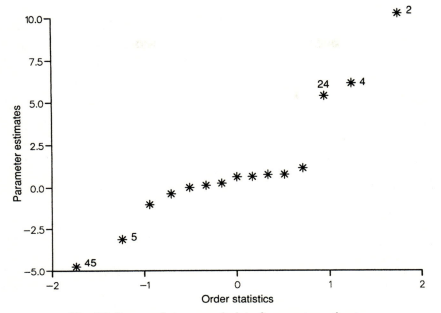

Fig. 8.9. Reactor data: normal plot of parameter estimates.

interactions. Table 8.8 is the analysis of variance obtained by pooling the remaining 10 degrees of freedom. This clearly shows the statistical significance of the five contrasts.

This example illustrates several points in the analysis of 2^m factorials and their fractions. Because the design is a 2^{5-1} fractional factorial, the five effects found are also estimating three four-factor and two three-factor interactions. On the grounds of simplicity it is reasonable to assume that the observed effects are due to the simpler set of parameters, namely the main effects and their interactions. These interactions are only between factors which are individually significant. It is very seldom that significant interactions are found between factors which are not themselves significant. If the variables x_j are scaled (§2.1), a return to the original units for a model containing a pure interaction will result in both main effect and interaction terms. Pure interactions are usually only meaningful when the units of the factors have a physical origin.

Figure 8.9 is a normal plot of parameter estimates. With a relatively small number of observations, a clearer idea of which part of the plot is straight is sometimes obtained from a half-normal plot such as Fig. 8.10. A disadvantage of plotting the absolute values of the parameter estimates is that information on signs is lost. Finally, once the important effects have been tentatively

Table 8.8. Reactor data

Source	Sum of squares	Analysis of variance with and without the component for interaction 35 pooled for error					
		Degrees of freedom	Mean square	F	Degrees of freedom	Mean square	F
2	1681	1	1681	239.3	1	1681	302.6
4	600.25	1	600.25	85.4	1	600.25	108.0
5	156.25	1	156.25	22.2	1	156.25	28.1
24	462.25	1	462.25	65.8	1	462.25	83.2
25	361.0	1	361.0	51.4	1	361.0	65.0
(35)	—				1	20.25	3.64
Residual (pooled)	70.25	10	7.025		9	5.556	
Total (corrected)	3331	15			15		
		$F_{1,10,0.1\%} = 21.04$			$F_{1,9,5\%} = 5.12$		

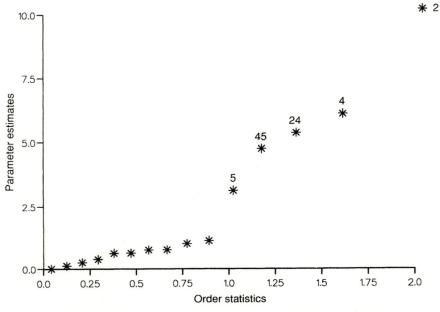

Fig. 8.10. Reactor data: half-normal plot of parameter estimates.

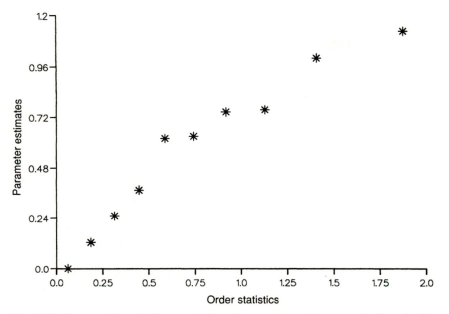

Fig. 8.11. Reactor data: half-normal plot of parameter estimates excluding the five most significant estimates.

identified, they can be removed from the plot, which can then be redrawn to check that only negligible effects remain. Figure 8.11 is such a half-normal plot of the 10 remaining effects, which does not show any striking departure from straightness.

It seems clear from this analysis that the important effects are 2, 4, 5, 24, and 25. The next largest estimate is for interaction 35. If it were not clear whether this term should also be included in the model, the analysis of variance of Table 8.8 could be extended as shown, to test the significance of this term, with a reduction to nine degrees of freedom for the estimate of error. Here the result is not significant, although the general procedure can be useful. For this example it would be surprising if interaction 35 were important, when the main effect of 3 is absent from the model.

A fuller analysis of the example would include residual plots, which are given by Box *et al.* (1978, p. 380) who, however, do not calculate an analysis of variance table. The residual plot does not suggest any inadequacies in the fitted model. However, the maximum theoretical value of the response (yield) is 100 per cent. The observed maximum is 95 per cent. Therefore it would be interesting to consider transformations of the form $(100 - y)^{\lambda}$ in an analysis similar to that of §8.2.

The motivation for use of the normal plots of parameter estimates was that it might not be known a priori which interactions were negligible. However, even if high-order interactions are estimable from a design and are believed to be negligible, use of plots like Figs 8.9 or 8.10 is recommended for the information on important and unimportant effects which may be revealed. The plots given by Box *et al.* (1978) have the x and y axes reversed compared with those shown here, a habit shared by some computer software and by Daniel (1958), who introduced the plots.

8.4 Further reading

Cox and Snell (1981) give a succinct, but thoughtful, introduction to the analysis of data. Atkinson (1985) contains much material on the analysis of transformations touched on in §8.2. Several introductory books on statistics include the analysis of 2^m factorials and their fractions. Hines and Montgomery (1990, pp. 413–48) give examples of detailed analyses including normal plots of effects, residual plots, and analyses of variance.

The notation of (8.1) and (8.8) is widely used in the analysis of data, for example by Atkinson (1985). The projection matrix H is called the 'hat' matrix since the vector of fitted values

$$\hat{y} = F\hat{\beta} = F(F^\mathrm{T}F)^{-1}F^\mathrm{T}y = Hy.$$

Then

$$\mathrm{var}\{\hat{y}(x_i)\} = \sigma^2 f^\mathrm{T}(x_i)\,(F^\mathrm{T}F)^{-1}f(x_i)$$

$$= \sigma^2 h_i \qquad\qquad (5.25)$$

where h_i is the ith diagonal element of H. If emphasis is on design, rather than analysis, as is the case in the remainder of this book, the importance of the design is stressed by writing

$$d(x, \xi) = \frac{N}{\sigma^2}\,\mathrm{var}\{\hat{y}(x)\}, \qquad\qquad (5.30)$$

whence

$$d(x, \xi) = Nh_i.$$

Part II
Theory and applications

9
Optimum design theory

9.1 Continuous and exact designs

In previous chapters we have presented ideas, mostly informal, about good experimental design, leading up to a discussion in Chapters 5 and 6 of some optimum designs for linear and quadratic regression in one factor. In this chapter we begin the second part of the book with a more formal discussion of the General Equivalence Theorem, which is the central result on which the optimum design of experiments depends. The theorem applies to a wide variety of design criteria, many of which are called by a letter of the alphabet, so that the subject is sometimes called 'alphabetic optimality'. These criteria are the subject of the next chapter. In order to avoid repetition, in this chapter we work with a general criterion Ψ which is to be minimized. A special case is D-optimality and its relationship to G-optimality. Because of the importance of D-optimality, Chapter 11 is devoted to this criterion, with examples of designs and of algorithms for constructing them. We begin this chapter with a more formal statement of the important distinction between exact and continuous designs and give the notation for them. On their first exposure to this material, readers may prefer to go straight to Chapter 11 after the end of this section.

In §6.3 we showed that the D-optimum design for the first-order model put half the trials at $x = -1$ and the other half at $x = +1$ when $\mathscr{X} = [-1, 1]$, provided that the number of trials N was even. For odd N the number is divided as equally as possible. Thus for increasing odd N there will be a sequence of exact designs, starting with a $2:1$ division of the trials when $N = 3$, which approaches the equal division for even N. The mathematical problem of finding the optimum design is simplified by considering only this asymptotic or continuous design, thus ignoring the constraint that the number of trials at any design point must be an integer.

Continuous designs are represented by the measure ξ over $\mathscr{X}$. If the design has trials at n distinct points in $\mathscr{X}$, we write

$$\xi = \begin{Bmatrix} x_1 & x_2 \dots x_n \\ w_1 & w_2 \dots w_n \end{Bmatrix} \tag{9.1}$$

where the first line gives the values of the factors at the design points with the w_i the associated design weights. Since ξ is a measure, $\int_{\mathscr{X}} \xi(\mathrm{d}x) = 1$ and $0 \leqslant w_i \leqslant 1$ for all i.

The D-optimum design for the first-order model found in §6.3 is then written as

$$\xi = \left\{ \begin{matrix} -1 & 1 \\ 1/2 & 1/2 \end{matrix} \right\}, \tag{9.2}$$

whereas the design for three trials, two of which are at the upper value of x, is

$$\xi = \left\{ \begin{matrix} -1 & 1 \\ 1/3 & 2/3 \end{matrix} \right\}.$$

If we wish to stress that a measure refers to an exact design, realizable in integers for a specific N, the measure is written

$$\xi_N = \left\{ \begin{matrix} x_1 & x_2 & \dots & x_n \\ r_i/N & r_2/N & \dots & r_n/N \end{matrix} \right\} \tag{9.3}$$

where r_i is the integer number of trials at x_i and $\Sigma_{i=1}^n r_i = N$.

In practice all designs are exact. For moderate N good exact designs can frequently be found by integer approximation to the optimum continuous measure ξ^*. Often, for simple models with p parameters, there will be p design points with equal weight $1/p$, so that the exact design with $N=p$ trials is optimum. However, if the design weights are not rational, it will not be possible to find an exact design which is identical with the continuous optimum design.

Example **9.1** Quadratic regression through the origin

The model is

$$\eta(x) = \beta_1 x + \beta_2 x^2$$

over the region $\mathscr{X} = [0, 1]$. The purpose is to estimate β_2 with minimum variance, which corresponds to checking whether curvature is present in data such as that for Example 1.1 (the desorption of carbon monoxide). As we show in §9.5, the optimum continuous measure is

$$\xi^* = \left\{ \begin{matrix} \sqrt{2}-1 & 1 \\ \sqrt{2}/2 & (2-\sqrt{2})/2 \end{matrix} \right\}. \tag{9.4}$$

Although good approximations to this design can be found for various values of N, there is no exact version of ξ^* for finite N.   □

Difficulties in finding exact designs usually arise when N is close to the number of support points of the optimum continuous design, leading to a poor approximation to ξ^*. Chapter 15 describes special algorithms for finding optimum exact designs and gives a variety of examples. In comparing exact designs it should be remembered that it is the value of the design criterion Ψ which is of importance, and not the closeness of ξ_N to ξ^*.

For an N-trial design the information matrix for β in the model $E(Y) = F\beta$ was defined in §5.3 as $F^T F$, where

$$F^T F = \sum_{i=1}^{N} f(x_i) f^T(x_i) \tag{9.5}$$

and $f^T(x_i)$ is the ith row of F. For the continuous design ξ, the information matrix is

$$M(\xi) = \int_{\mathscr{X}} m(x)\xi(\mathrm{d}x) = \int_{\mathscr{X}} f(x)f^T(x)\xi(\mathrm{d}x)$$

$$= \sum_{i=1}^{n} f(x_i)f^T(x_i)w_i. \tag{9.6}$$

The last form in (9.6) is summed over the n design points which, because of the presence of the weights w_i, becomes a scaled version of (9.5) for the exact design ξ_N, i.e.

$$M(\xi_N) = \frac{F^T F}{N}.$$

The variance of the predicted response for an N-trial design was given in (5.25) as

$$\mathrm{var}\{\hat{y}(x)\} = \sigma^2 f^T(x) (F^T F)^{-1} f(x).$$

For continuous designs the standardized variance of the predicted response is

$$d(x, \xi) = f^T(x)M^{-1}(\xi)f(x), \tag{9.7}$$

a function of both the design ξ and the point at which the prediction is made. If the design is exact,

$$d(x, \xi_N) = f^T(x)M^{-1}(\xi_N)f(x) = \frac{N \, \mathrm{var}\{\hat{y}(x)\}}{\sigma^2},$$

the standardized variance introduced in (5.30).

9.2 The General Equivalence Theorem

In the theory for continuous designs we consider minimization of the general measure of imprecision $\Psi\{M(\xi)\}$. Under very mild assumptions, the most important of which are the compactness of $\mathscr{X}$ and the convexity and differentiability of Ψ, designs which minimize Ψ also satisfy a second criterion.

One example is D-optimality, in which $\Psi\{M(\xi)\}=\log|M^{-1}(\xi)|=-\log|M(\xi)|$ so that the determinant of the information matrix $M(\xi)$ is maximized. Taking the logarithm of the determinant leads to minimization of a convex function, so that any minimum found will certainly be global rather than local. Continuous designs which are D-optimum are also G-optimum, i.e. they minimize the maximum over $\mathscr{X}$ of the variance (9.7).

The General Equivalence Theorem can be viewed as an application of the result that the derivatives are zero at a minimum of a function. However, the function depends on the measure ξ through the information matrix $M(\xi)$. Let the measure $\bar{\xi}$ put unit mass at the point x and let the measure ξ' be given by

$$\xi'=(1-\alpha)\xi+\alpha\bar{\xi}.$$

Then, from (9.6),

$$M(\xi')=(1-\alpha)M(\xi)+\alpha M(\bar{\xi}). \tag{9.8}$$

Accordingly, the derivative of Ψ in the direction $\bar{\xi}$ is

$$\phi(x,\xi)=\lim_{\alpha\to0^+}\frac{1}{\alpha}\left[\Psi\{(1-\alpha)M(\xi)+\alpha M(\bar{\xi})\}-\Psi\{M(\xi)\}\right]. \tag{9.9}$$

The General Equivalence Theorem then states the equivalence of the following three conditions on ξ^*:

(1) the design ξ^* minimizes $\Psi\{M(\xi)\}$;

(2) the minimum of $\phi(x,\xi^*)\geqslant0$;

(3) the derivative $\phi(x,\xi^*)$ achieves its minimum at the points of the design. (9.10)

This theorem provides methods for the construction and checking of optimum designs. However, it says nothing about n, the number of support points of the design. A bound on this number can be obtained from the nature of $M(\xi)$ which is a $p\times p$ symmetric matrix. Because of the additive nature of information matrices (9.6), the information matrix of any design can be represented as a weighted sum of, at most, $p(p+1)/2$ information matrices $m(\bar{\xi}_i)$ where $\bar{\xi}_i$ puts unit weight at the design point x_i. Thus, even if an optimum design is found which contains more than $p(p+1)/2$ points, a design with the same information matrix, and so the same optimum value of $\Psi\{M(\xi)\}$, can be found which has support at no more than $p(p+1)/2$ points. Usually optimum designs contain fewer design points. For many D-optimum designs, the design contains p points, each with weight $1/p$.

The bound on the number of design points depends on the linear structure of $M(\xi)$ and so holds for any criteron which is a function of a single information matrix, as are nearly all the criteria discussed in this book. However, the criteria for the Bayesian designs of Chapter 19 are non-linear functions of several information matrices, as are some of the composite design criteria of Chapter 21. As examples show, the number of support points for these optimum designs is not bounded by $p(p+1)/2$. However, the optimum designs do satisfy the General Equivalence Theorem (9.10).

Example 9.2 Quadratic regression (Example 4.2 continued)

The D-optimum continuous design for the quadratic model in one variable with $\mathcal{X} = [-1, 1]$ is

$$\xi^* = \left\{ \begin{matrix} -1 & 0 & 1 \\ 1/3 & 1/3 & 1/3 \end{matrix} \right\}. \tag{9.11}$$

The claim that this design is D-optimum can be checked using the General Equivalence Theorem.

For D-optimality, when $\Psi\{M(\xi)\} = -\log|M(\xi)|$, the derivative function is

$$\phi(x, \xi) = p - d(x, \xi) \tag{9.12}$$

where $d(x, \xi)$ is the standardized variance function (9.7). From Condition 2 of (9.10), that $\phi(x, \xi^*) \geq 0$, it follows that

$$d(x, \xi^*) \leq p. \tag{9.13}$$

For the continuous design (9.11), $d(x, \xi^*)$ is the quadratic given by (5.33). The plot in Fig. 9.1 shows that the maximum value of $d(x, \xi^*)$ is indeed p, in this case 3, so that the design is D-optimum, as claimed. Further, Condition 3 of the theorem is satisfied, since these maxima occur at the points of support of the design.   □

Plots, similar to Fig. 9.1, of the derivative functions of the criteria to be described in Chapter 10 can be used in the same manner to explore the properties of designs and to check the optimality of continuous designs, particularly when $\mathcal{X}$ is of low dimension. However, it is essential that $\mathcal{X}$ is searched thoroughly, as is shown by the following example, again for D-optimality.

Example 9.3 Simple regression (Example 5.1 continued)

For simple regression consider the design

$$\xi = \left\{ \begin{matrix} -a & a \\ 1/2 & 1/2 \end{matrix} \right\}$$

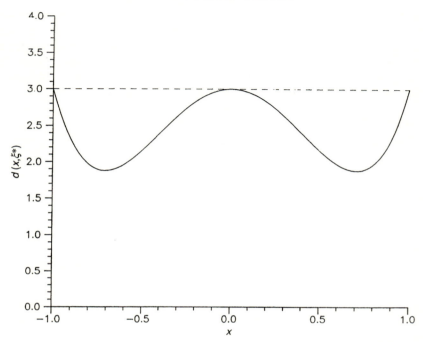

Fig. 9.1. Example 9.2: quadratic regression. Variance function $d(x, \xi^*)$ for the D-optimum continuous design. The maximum value of 3 occurs at the design points.

with $0 < a < 1$ and $\mathscr{X} = [-1, 1]$. To check the D-optimality of this design calculate

$$d(x, \xi) = 1 + x^2/a^2. \qquad (9.14)$$

Then at the points of the design $x = a$ and $d(x, \xi) = 2$, the number of parameters in the model. However, the design, as we know from §6.3, is not D-optimum. The maximum of (9.14) is $1 + 1/a^2$, which is greater than 2, at $x = \pm 1$. This example illustrates that it is necessary to check the value of the derivative function not only at the design points but over the whole of $\mathscr{X}$.   □

9.3 Exact designs and the General Equivalence Theorem

The General Equivalence Theorem holds for continuous designs represented by the measure ξ. In general, it does not hold for exact designs. For D-optimality the implication is that there will be some values of N for which one design will be D-optimum and another G-optimum. Examples were given in §5.3. We now explore one of these more fully and calculate a G-optimum design.

***Example* 9.4** Quadratic regression, $N=4$

When $\mathscr{X}=[-1, 1]$, the D-optimum continuous design (9.11) puts weight $1/3$ at the three points $x=-1$, 0, and 1. This division provides exact optimum designs when N is a multiple of 3, but not otherwise. For $N=4$, the exact D-optimum design consists of replicating any one of the design points to give a design for which $|M(\xi_4^*)|=1/8$. The General Equivalence Theorem cannot be used to prove that this design is D-optimum. Suppose that the centre point is replicated, giving the design

$$\xi_4^*=\begin{Bmatrix} -1 & 0 & 1 \\ 1/4 & 1/2 & 1/4 \end{Bmatrix}. \tag{9.15}$$

Then, as Fig. 9.2 shows, the local maxima of the variance function are $d(-1, \xi_4^*)=d(1, \xi_4^*)=4$, with $d(0, \xi_4^*)=2$, so that the maximum value is greater than p, which is 3. In fact, there is no easy demonstration that (9.15) yields the D-optimum exact design. In most examples computer searches of the kind described in Chapter 15 have to suffice.

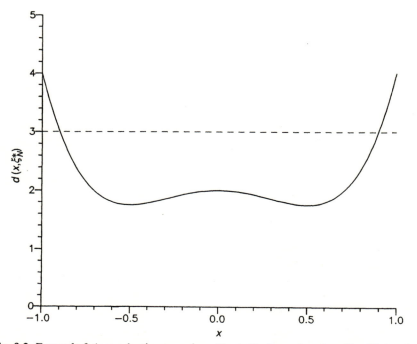

Fig. 9.2. Example 9.4: quadratic regression, $N=4$. Variance function $d(x, \xi_N^*)$ for the D-optimum exact design.

However, it is possible, in this example, to find the G-optimum exact design for $N=4$. Figure 9.2 shows that the design (9.15) concentrates too much design weight in the centre of the region. The symmetrical four-point design

$$\xi_4 = \left\{ \begin{matrix} -1 & -a & a & 1 \\ 1/4 & 1/4 & 1/4 & 1/4 \end{matrix} \right\} \tag{9.16}$$

can have the value of a chosen so that $d(x, \xi)$ has the same value at $x=0$ and $x=1$. This requirement yields $a^2 = \sqrt{5}-2$, when $d(0, \xi)=d(1, \xi)=3.618$.

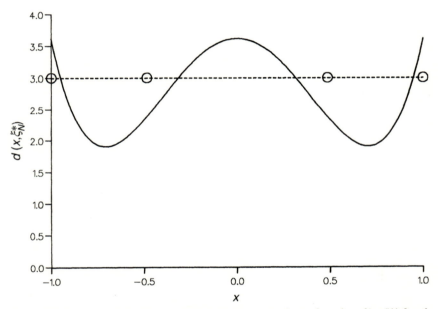

Fig. 9.3. Example 9.4: quadratic regression, $N=4$. Variance function $d(x, \xi_N^*)$ for the G-optimum exact design: ○ design points.

Figure 9.3 shows that this is the maximum value over $\mathscr{X}$ of $d(x, \xi)$. Although this value is greater than 3, this is the exact G-optimum design for $N=4$. A feature which complicates the search for G-optimum exact designs is that the maxima of $d(x, \xi_N)$ do not necessarily occur at the design points. A last property of this design is that $|M(\xi_4)|=0.0902<0.125$, the value for the D-optimum designs such as (9.15).   □

In most examples computer searches are needed to find exact designs for specific N. Because ξ_N is discrete, the search for the design minimizing $\Psi\{M(\xi_N)\}$ is a combinatorial optimization problem. For exact G-optimum designs the difficulty of the optimization is compounded by the need to search over $\mathscr{X}$ to identify the maximum of $d(x, \xi_N)$ for each trial design measure ξ_N. Sometimes, if an exact design is required which is close to being G-optimum, the search for this second maximum is replaced by evaluation of $d(x, \xi_N)$ over a coarse grid in $\mathscr{X}$. This method can also be used for the comparison of designs which behave almost equally well according to some other criterion, for example exact D-optimality.

9.4 Algorithms for continuous designs and the General Equivalence Theorem

We return to consideration of algorithms for finding continuous measures ξ which minimize $\Psi\{M(\xi)\}$. From the General Equivalence Theorem (9.10) it follows that at the optimum design all gradients $\phi(x, \xi^*)$ are non-negative. Away from the optimum there will be some direction in which $\phi(x, \xi) < 0$. If, as in §9.2, we let the measure $\bar{\xi}_k$ put unit mass at the point x_k, chosen so that $\phi(x, \xi_k) < 0$, the algorithm

$$\xi_{k+1} = (1 - \alpha_k)\xi_k + \alpha_k \bar{\xi}_k \tag{9.17}$$

will lead to a decrease in Ψ provided that the step length α_k for the kth step is chosen to be sufficiently small.

The algorithms given by (9.17) are a family of descent algorithms. Usually $\bar{\xi}_k$ is chosen so that x_k is the point for which $\phi(x_k, \xi_k)$ is a minimum, when the algorithm becomes one of steepest descent. Although it is possible to search in this direction and to find α_k in (9.17) for which $\Psi\{M(\xi_{k+1})\}$ is minimized, a fixed sequence of step lengths is often used. In the original algorithms for the construction of D-optimum designs $\alpha_k = (k+1)^{-1}$. Let

$$\bar{d}(\xi) = \max_{x \in \mathscr{X}} d(x, \xi). \tag{9.18}$$

Then, from (9.12), the steepest descent algorithm for D-optimality will successively add mass to the design measure corresponding to the point where $\bar{d}(\xi)$ is attained. The choice of weights $\alpha_k = (k+1)^{-1}$ yields an algorithm in which the optimum design is constructed by the sequential addition of trials at the points where $d(x, \xi_k)$ is a maximum. A numerical example of such an algorithm is given in Chapter 11.

The convergence of these first-order algorithms is often slow. However, ultimate convergence is usually not required. Starting from an arbitrary design, the algorithm typically adds mass at a restricted number of conditions which suggest the support of the optimum design. Once this pattern becomes clear, the arbitrary starting design can be rejected and the algorithm restarted. Alternatively, a second-order optimization algorithm can be employed to minimize $\Psi\{M(\xi)\}$. The advantage of the first-order algorithm is that it reduces the search for an optimum design to a sequence of optimizations in m dimensions.

9.5 Function optimization and continuous designs

It is sometimes possible to find an optimum continuous design by algebraic minimization. We begin this section by giving an example, before considering some transformations which are helpful in the construction of optimum designs by numerical optimization.

Example 9.1 Quadratic regression through the origin

The design for estimating β_2 with minimum variance is given in (9.4). To find this optimum design analytically consider the measure

$$\xi = \left\{ \begin{matrix} a & 1 \\ w & 1-w \end{matrix} \right\}. \tag{9.19}$$

It is clear that one point of support of the design must be at $x = 1$, since any design for which this is not the case can be scaled up, with a consequent increase in all elements of the information matrix.

For quadratic regression through the origin it follows from (9.6) that the information matrix is

$$M(\xi) = \begin{bmatrix} \int x^2 \xi(\mathrm{d}x) & \int x^3 \xi(\mathrm{d}x) \\ \int x^3 \xi(\mathrm{d}x) & \int x^4 \xi(\mathrm{d}x) \end{bmatrix}.$$

For the measure (9.19) this becomes

$$M(\xi) = \begin{bmatrix} 1-w+wa^2 & 1-w+wa^3 \\ 1-w+wa^3 & 1-w+wa^4 \end{bmatrix}. \tag{9.20}$$

From (9.20),

$$|M(\xi)| = (1-w+wa^2)(1-w+wa^4) - (1-w+wa^3)^2$$
$$= w(1-w)a^2(1-a)^2.$$

Thus, from (9.20), the (2, 2) element of $M^{-1}(\xi)$ is

$$\frac{1-w+wa^2}{w(1-w)a^2(1-a)^2}. \tag{9.21}$$

The design for which var $\hat{\beta}_2$ is minimized will therefore minimize (9.21). Differentiation of (9.21) followed by setting the derivatives equal to zero leads to the optimum design (9.4). ☐

The design for which the variance of a single-parameter estimate is minimized is a special case of D_s-optimality, in which the generalized variance of a subset of the parameters is minimized. The corresponding equivalence theorem is given in Chapter 10.

If it were not possible to maximize (9.21) analytically, a numerical method, such as a quasi-Newton algorithm, could be used to find the minimum of this function of a and w. However, the minimization would be subject to the constraints $0 \leqslant w \leqslant 1$ and $0 \leqslant a \leqslant 1$. Although general methods for constrained optimization could be used, it is often possible, for straightforward design problems, to use transformations which lead to an unconstrained search in a suitably defined space.

Suppose that x is constrained to be non-negative, i.e. $0 \leqslant x < \infty$. For example, x could be a quantitative factor such as time or dose of a drug. Then the transformation

$$x = e^z \qquad (-\infty < z < \infty) \tag{9.22}$$

ensures that the constraint on x is satisfied for any value of z. A disadvantage of (9.22) is that $x = 0$ corresponds to $z = -\infty$. A preferable transformation is to put

$$x = z^2 \qquad (-\infty < z < \infty) \tag{9.23}$$

where $x = 0$ is in the centre of the range of z. If (9.23) is used with a numerical search algorithm, the steps in z should not be so large that the algorithm oscillates between positive and negative values of z.

For the cubic design region with $-1 \leqslant x_i \leqslant 1$ $(i = 1, 2, \ldots, m)$ the appropriate transformation is to put

$$x_i = \sin z_i \qquad (i = 1, 2, \ldots, m) \tag{9.24}$$

or equivalently $x_i = \cos x_i$. Searching in the unconstrained space of the z_i leads to a constrained search in x space. Again, care is needed in the choice of step length for the search.

If $\mathscr{X}$ is the m-dimensional unit sphere centred at the origin, the transformation is more complicated. One possibility, using polar co-ordinates, is

$$x_1 = \sin z_1$$

$$x_2 = \sin z_2 \cos z_1$$

$$\vdots$$

$$x_i = \sin z_i \prod_{j=1}^{i-1} \cos z_j \qquad (i = 2, \ldots, m). \tag{9.25}$$

As a result of (9.25) the x_i satisfy the constraints

$$-1 \leqslant x_i \leqslant 1 \qquad (i = 1, \ldots, m)$$

and

$$\sum_{i=1}^{m} x_i^2 \leqslant 1 \tag{9.26}$$

for all z_i.

The search for an optimum design measure involves constraints not only on the x's arising from $\mathscr{X}$ but also on the design weights w_i. In place of (9.26) these constraints are

$$0 \leqslant w_i \leqslant 1 \qquad \text{and} \qquad \sum_{i=1}^{n} w_i = 1. \tag{9.27}$$

The appropriate transformation is

$$w_1 = \sin^2 z_1$$
$$w_2 = \sin^2 z_2 \cos^2 z_1$$
$$\vdots$$
$$w_i = \sin^2 z_i \prod_{j=1}^{i-1} \cos^2 z_j \qquad (i = 2, \ldots, n-1)$$
$$\vdots$$
$$w_n = \prod_{j=1}^{n-1} \cos^2 z_j. \tag{9.28}$$

It should be noted that in (9.28) there are only $n-1$ values of z since the weights sum to unity. The transformation (9.28) can also be used for searching over the design region for mixture experiments.

It is our experience that these transformations, combined with a quasi-Newton method using numerical derivatives, provide a powerful method for finding optimum continuous designs. If the structure of the design is known approximately, convergence of the algorithm can be speeded up by choosing narrower intervals than those given by (9.24). If these are centred, for x_i, on the value x_{i0}, the transformation is

$$x_i = x_{i0} + d_i \sin z_i,$$

with $d_i < 1$.

9.6 Further reading

The original equivalence theorem between D- and G-optimality is due to Kiefer and Wolfowitz (1960). A much more general statement at appreciable length is given by Kiefer (1974). Whittle (1973) provides a succinct proof of a very general version of the theorem. Silvey (1980, Chapter 3) gives a careful discussion and proof of the theorem of §9.2. The argument for the bound on the number of support points of the design depends upon Carathéodory's Theorem for the representation of an arbitrary design matrix as a convex combination of unitary design matrices (Silvey 1980, Appendix 2). Wu and Wynn (1978) discuss the convergence of first-order algorithms for design measures given by (9.17). Convergence of the algorithm for D-optimum designs was proved by Wynn (1972). However, convergence is of the information matrix $M(\xi)$ to $M(\xi^*)$. It is possible to construct examples where convergence is to a design with more than $p(p+1)/2$ design points (Chan 1990). Such behaviour is of mathematical, rather than practical, importance.

10
Criteria of optimality

10.1 *A*-, *D*- and *E*-optimality

In this chapter we describe the more important special cases of the design criterion $\Psi\{M(\xi)\}$ of Chapter 9 and give the derivative functions $\phi(x, \xi)$ for the particular General Equivalence Theorems analogous to (9.10).

The most important design criterion in applications is that of *D*-optimality, in which the generalized variance, or its logarithm $-\log|M(\xi)|$, is minimized. The relationship between this criterion and *G*-optimality was extensively discussed in Chapter 9. Two other criteria which have a statistical interpretation in terms of the information matrix $M(\xi)$ are *A*- and *E*-optimality. In *A*-optimality $\text{tr}\{M^{-1}(\xi)\}$, the average variance of the parameter estimates, is minimized. In *E*-optimality the variance of the least well-estimated contrast $a^{\text{T}}\beta$ is minimized subject to the constraint $a^{\text{T}}a = 1$. Thus the *E* in the name of this criterion stands for extreme.

D-optimality was motivated in Chapter 6 by reference to the ellipsoidal confidence regions for the parameters of the linear model. A *D*-optimum design minimizes the content of this confidence region and so minimizes the volume of the ellipsoid. However, other properties of the confidence region may be of interest. A long thin ellipsoid orientated along or close to the parameter axes will result in comparatively poor estimation of one or more parameters, the variances of which will then be unduly large. If the ellipsoid is oriented at an appreciable angle to the axes, the variance of each individual parameter estimate may be satisfactorily small; however, owing to the correlations between the estimates, there will be contrasts in the parameters, corresponding to the directions of the long axes of the ellipsoid, which will be imprecisely estimated. It is these situations which are addressed by *A*- and *E*-optimality respectively.

The above ideas can be put more formally by considering the eigenvalues $\lambda_1, \ldots, \lambda_p$ of $M(\xi)$. The eigenvalues of $M^{-1}(\xi)$ are then $1/\lambda_1, \ldots, 1/\lambda_p$ and are proportional to the squares of the lengths of the axes of the confidence ellipsoid. In terms of these eigenvalues the three criteria are as follows.

A Minimize the sum (or average) of the variances of the parameter estimates:

$$\min \sum_{i=1}^{p} \frac{1}{\lambda_i}.$$

D Minimize the generalized variance of the parameter estimates:

$$\min \prod_{i=1}^{p} \frac{1}{\lambda_i}.$$

E Minimize the variance of the least well-estimated contrast $a^T\beta$ with $a^Ta = 1$:
$\min \max(1/\lambda_i)$.

All three can be regarded as special cases of the more general criterion of choosing designs to minimize

$$\Psi_k(\xi) = \left(p^{-1} \sum_{i=1}^{p} \lambda_i^{-k} \right)^{1/k} \qquad (0 \leqslant k < \infty).$$

For *A-*, *D-*, and *E*-optimality the values of k are $1, 0,$ and ∞ respectively when the limiting operations are properly defined. Kiefer (1975) uses this family to study the variation in structure of the optimum design as the optimality criterion changes in a smooth way.

In order to state the equivalence theorems for these and other criteria it is convenient to rewrite the derivative (9.9) as

$$\phi(x, \xi) = \Delta(\xi) - \psi(x, \xi),$$

where

$$\Delta(\xi) = -\operatorname{tr} M \frac{\partial \Psi}{\partial M} \qquad \psi(x, \xi) = -f^T(x) \frac{\partial \Psi}{\partial M} f(x)$$

$$\Psi = \Psi\{M(\xi)\} \qquad M = M(\xi). \qquad (10.1)$$

The functions for *A-*, *D-*, and *E*-optimality are given in Table 10.1, together with the functions for some other criteria. In Table 10.1 *A*-optimality is the special case of linear optimality with $A = I$, the identity matrix.

An advantage of *D*-optimality is that the optimum designs for quantitative factors do not depend upon the scale of the variables. Linear transformations leave the *D*-optimum design unchanged, which is not in general the case for *A*- and *E*-optimum designs. In principle this is a serious drawback to the other two criteria: it seems undesirable that an optimum design should depend upon whether a factor is measured in inches or centimetres. However, as we have seen, it is customary to work with scaled variables x_j, rather than with the unscaled u_j, so that the units of measurement are irrelevant. For designs with all factors qualitative, such as block designs, the problem of scale does not arise and *A*- and *E*-optimum designs are frequently employed. One reason is the relative ease of construction of such designs in non-standard situations.

Table 10.1. Functions appearing in the General Equivalence Theorem (10.1) for a variety of optimality criteria

Criterion	Ψ	ψ	Δ
D	$\log\|M^{-1}\|$	$f(x)^T M^{-1} f(x)$	p
Linear	$\operatorname{tr} A M^{-1}$	$f(x)^T M^{-1} A M^{-1} f(x)$	$\operatorname{tr} A M^{-1}$
E	$\min \lambda_i(M) = \lambda_{\min}$	$f(x)^T r r^T f(x)^*$	$\lambda_{\min}$
Generalized D	$\log\|A^T M^{-1} A\|$	$f(x)^T M^{-1} A \{A^T M^{-1} A\}^{-1}$ $A^T M^{-1} f(x)$	$s = \operatorname{rank} A$
Generalized G	$\max_{x \in Z} w(x) d(x, \xi) = C(\xi)$	$f(x)^T M^{-1} \int_Z w(x) f(x) f(x)^T \, dx$ $M^{-1} f(x)$	$C(\xi)$
q	$q^{-1} \operatorname{tr} M^{-q}$	$f(x)^T M^{-q-1} f(x)$	$\operatorname{tr} M^{-q}$

*r is the eigenvector for $\lambda_{\min}$.

D-optimum designs are more readily constructed for experiments with quantitative factors. We now consider several useful extensions to D-optimality.

10.2 D_A-optimality

Sometimes interest is not in all p parameters, but only in s linear combinations of β which are the elements of $A^T \beta$, where A is $p \times s$ of rank $s < p$. The covariance matrix for these linear combinations is $A^T M^{-1}(\xi) A$. If $s = 1$, designs minimizing the three criteria of §10.1 for this covariance matrix all reduce to the c-optimum designs of §10.5 in which the variance of the estimated linear combination is minimized. When $s > 1$, the A-optimum design minimizing the trace of $A^T M^{-1}(\xi) A$ is an example of the linear optimum designs of §10.6. In this section we define D-optimum designs minimizing

$$\Psi\{M(\xi)\} = \log|A^T M^{-1}(\xi) A|.$$

To emphasize the dependence of the design on the matrix of coefficients A, this criterion is called D_A-optimality (Sibson 1974). The analogue of the variance function $d(x, \xi)$ (9.7) is

$$d_A(x, \xi) = f^T(x) M^{-1}(\xi) A \{A^T M^{-1}(\xi) A\}^{-1} A^T M^{-1}(\xi) f(x). \qquad (10.2)$$

If

$$\bar{d}_A(\xi) = \sup_{x \in \mathscr{X}} d_A(x, \xi),$$

then $\bar{d}_A(\xi^*) = s$, where now ξ^* is the continuous D_A-optimum design. When the design is optimum the maxima of this function again occur at the points of support of the design. One application of D_A-optimum design, described in §22.4, is in the allocation of treatments in clinical trials. We now consider an important special case of D_A-optimality.

10.3 D_s-optimality

D_s-optimum designs are appropriate when interest is in estimating a subset of s of the parameters as precisely as possible. Let the terms of the model be divided into two groups:

$$E(Y) = f^T(x)\beta = f_1^T(x)\beta_1 + f_2^T(x)\beta_2 \qquad (10.3)$$

where the β_1 are the s parameters of interest. The $p - s$ parameters β_2 are then treated as nuisance paremeters. One example is when β_1 corresponds to the experimental factors and β_2 corresponds to the parameters for the blocking factors. A second example is when experiments are designed to check the form of a model. Then, to be consistent with the notation of this section, the tentative model with terms $f_2(x)$ is embedded in a more general model by the addition of the terms $f_1(x)$. In order to test whether the simpler model is adequate, precise estimation of β_1 is required. A fuller description of this procedure can be found in §21.5. Examples of D_s-optimum designs for blocking are given in Chapter 14.

To obtain expressions for the design criterion and related variance function, partition the information matrix as

$$M(\xi) = \begin{bmatrix} M_{11}(\xi) & M_{12}(\xi) \\ M_{12}^T(\xi) & M_{22}(\xi) \end{bmatrix}.$$

The covariance matrix for the least squares estimates of β_1 is $M^{11}(\xi)$, the $s \times s$ left upper submatrix of $M^{-1}(\xi)$. It is easy to verify, from results on the inverse of a partitioned matrix, that

$$M^{11}(\xi) = \{M_{11}(\xi) - M_{12}(\xi)M_{22}^{-1}(\xi)M_{12}^T(\xi)\}^{-1}.$$

The D_s-optimum design for β_1 accordingly maximizes the determinant

$$|M_{11}(\xi) - M_{12}(\xi)M_{22}^{-1}(\xi)M_{12}^T(\xi)| = \frac{|M(\xi)|}{|M_{22}(\xi)|}. \qquad (10.4)$$

The right-hand side of (10.4) leads to the expression for the variance

$$d_s(x, \xi) = f^T(x)M^{-1}(\xi)f(x) - f_2^T(x)M_{22}^{-1}(\xi)f_2(x). \qquad (10.5)$$

For the D_s-optimum design ξ^*

$$d_s(x, \xi^*) \leqslant s \qquad (10.6)$$

with equality at the points of support of the design. These results follow from those for D_A-optimality by taking $A = (I_s, 0)$, where I_s is the $s \times s$ identity matrix.

A mathematical difficulty that arises with D_s-optimum designs, and with some other designs such as the c-optimum designs of §10.5, is that $M(\xi^*)$ may be singular. As a result only certain linear combinations or subsets of the parameters may be estimable. The consequent difficulties in the proof of equivalence theorems are discussed by, for example, Silvey (1980, p. 25) and Pazman (1986, p. 122). The problem is avoided in the numerical construction of designs by regularizing the information matrix through the addition of a small multiple of the identity matrix. That is, we let

$$M_\varepsilon(\xi) = M(\xi) + \varepsilon I \qquad (10.7)$$

for ε small, but large enough to permit inversion of $M_\varepsilon(\xi)$ (Vuchkov 1977). The first-order design algorithm (9.17) can then be used for numerical calculation of optimum designs.

An example is given in §18.3, where designs are found for various properties of a non-linear model, for which the information matrix is singular.

We conclude this section with two examples of D_s-optimum designs.

***Example* 10.1** Quadratic regression (Example 5.2 continued)

The D-optimum continuous design for quadratic regression in one variable with $\mathcal{X} = [-1, 1]$ is given in (9.11). It consists of weight $1/3$ where $x = -1, 0$, and 1. The D_s-optimum design for the coefficient of the quadratic term is

$$\xi^* = \begin{bmatrix} -1 & 0 & 1 \\ 1/4 & 1/2 & 1/4 \end{bmatrix}. \qquad (10.8)$$

The points of support of the D- and D_s-optimum designs are thus the same, but the weights are different.

To verify that (10.8) is D_s-optimum we use the condition on the variance $d_s(x, \xi^*)$ given by (10.6). Since interest is in the quadratic term, the coefficients of the constant and linear terms are nuisance parameters. For the design (10.8) the information matrix thus partitions as

$$M(\xi^*) = \begin{bmatrix} \Sigma x^4 & \vdots & \Sigma x^2 & 0 \\ \cdots & \vdots & \cdots & \cdots \\ \Sigma x^2 & \vdots & 1 & 0 \\ 0 & \vdots & 0 & \Sigma x^2 \end{bmatrix} = \begin{bmatrix} 1/2 & \vdots & 1/2 & 0 \\ \cdots & \vdots & \cdots & \cdots \\ 1/2 & \vdots & 1 & 0 \\ 0 & \vdots & 0 & 1/2 \end{bmatrix}.$$

Then

$$M^{-1}(\xi^*) = \begin{bmatrix} 4 & -2 & 0 \\ -2 & 2 & 0 \\ 0 & 0 & 2 \end{bmatrix}$$

and

$$M_{22}^{-1}(\xi^*) = \begin{bmatrix} 1 & 0 \\ 0 & 2 \end{bmatrix}.$$

So, from (10.5),

$$d_s(x, \xi^*) = 4x^4 - 2x^2 + 2 - (2x^2 + 1)$$

$$= 4x^4 - 4x^2 + 1. \tag{10.9}$$

This quartic equals unity at $x = -1, 0$, or 1. These are the three maxima over $\mathscr{X}$ since, for $-1 < x < 1$, $x^4 \leqslant x^2$ with equality only at $x = 0$. Thus (10.8) is the D_s-optimum design for the quadratic term.   □

Example 10.2 Quadratic regression through the origin (Example 9.1 continued)

The D_s-optimum design of the previous example provides an estimate of the quadratic term with minimum variance. This design would be appropriate for testing whether the quadratic term should be added to a first-order model. Similarly, the design given by (9.4) is appropriate for checking curvature when the regression line passes through the origin.

The design (9.4) was calculated in §9.5 by direct minimization of $\text{var}(\hat{\beta}_2)$. To show that this design is D_s-optimum for β_2 we calculate $d_s(x, \xi^*)$. Substitution of (9.4) in (10.5) yields the numerical expression

$$d_s(x, \xi^*) = 25.73x^2 - 56.28x^3 + 33.97x^4 - 2.42x^2$$

$$= 23.31x^2 - 56.28x^3 + 33.97x^4.$$

Figure 10.1 is a plot of this curve, which does indeed have a maximum value of unity at the design points, which are marked by circles. The difference between Fig. 10.1 and the variance curves of Chapter 6 is interesting. Here, because the model passes through the origin, the variance of prediction must be zero when $x = 0$.   □

These two examples illustrate some of the properties of D_s-optimum designs. However, in both cases, $s = 1$. The designs are therefore optimum by several criteria discussed in this chapter, for example c-optimality in which the variance of a linear combination of the parameters is minimized. However, for

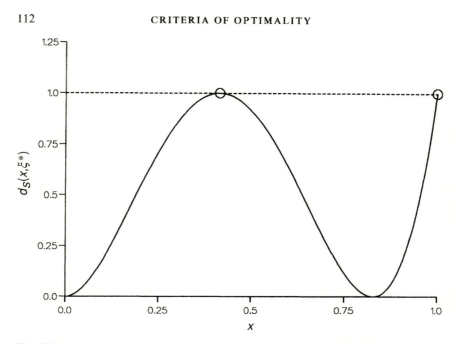

Fig. 10.1. Example 10.2: quadratic regression through the origin. Variance function $d_s(x, \xi^*)$ for the D_s-optimum design for β_2.

$s \geqslant 2$, D_s-optimum designs will in general be different from those satisfying other criteria.

10.4 Generalized D-optimality

The D_A-optimum design criterion of §10.2 is one extension of D-optimality. A further extension is to consider designs minimizing

$$\sum_{i=1}^{h} w_i \log|A_i^T M_i^{-1}(\xi)A_i| \tag{10.10}$$

for fixed non-negative weights w_i. This criterion, called S-optimality by Läuter (1976), is a linear combination of D_A-optimum criteria for various models and weights. It is thus another convex design criterion. The equivalence theorem states that

$$\sum_{i=1}^{h} w_i f_i^T(x)M_i^{-1}(\xi^*)A_i\{A_i^T M_i^{-1}(\xi)^* A_i\}^{-1} A_i^T M_i^{-1}(\xi^*)f_i(x) \leqslant \sum_{i=1}^{h} w_i s_i, \tag{10.11}$$

where s_i is the rank of A_i. This criterion has been used to design experiments for the simultaneous estimation of parameters in a variety of models, whilst estimating subsets for discrimination between models (Atkinson and Cox 1974). In the quadratic examples of the previous section w would reflect the balance between estimation of all the parameters in the model and the precise estimation of β_2. This is one of the topics discussed in more detail in Chapter 21.

10.5 *c*-optimality

In *c*-optimality interest is in estimating the linear combination of the parameters $c^T\beta$ with minimum variance. The design criterion to be minimized is thus

$$\text{var } c^T\hat{\beta} \propto c^T M^{-1}(\xi) \, c \qquad (10.12)$$

where c is $p \times 1$. The equivalence theorem states that, for the optimum design,

$$\{f^T(x)M^{-1}(\xi^*)c\}^2 \leqslant c^T M^{-1}(\xi^*) \, c \qquad (10.13)$$

for $x \in \mathscr{X}$.

Examples of *c*-optimum designs for a non-linear model are given in Chapter 18. A disadvantage of *c*-optimum designs is that they are often singular. For example, if c is taken to be $f(x_0)$, for a specific $x_0 \in \mathscr{X}$, the criterion reduces to minimizing the variance of prediction of the response at x_0. One way to achieve this is to perform all trials at x_0, a singular optimum design which is therefore non-informative about any other aspects of the model and data. The linear optimality criterion of the next section is an extension of *c*-optimality to designs for two or more contrasts, which can avoid the difficulties of singular designs.

10.6 Linear optimality (*C*- and *L*-optimality)

Let L be a $p \times q$ matrix of coefficients. Then minimization of the criterion function

$$\text{tr}\{M^{-1}(\xi)L\} \qquad (10.14)$$

leads to a linear, or *L*-optimum, design. The linearity in the name of the criterion is thus in the elements of the covariance matrix $M^{-1}(\xi)$. We now consider the relationship of this criterion to the others of this chapter.

If L is of rank $s \leqslant q$ it can be expressed in the form $L = AA^T$ where A is a $p \times s$ matrix of rank s. Then

$$\text{tr}\{M^{-1}(\xi)L\} = \text{tr}\{M^{-1}(\xi)AA^T\} = \text{tr}\{A^T M^{-1}(\xi)A\}. \qquad (10.15)$$

This form stresses the relationship with the D_A-optimum designs of §10.2, where the determinant, rather than the trace, of $A^T M^{-1}(\xi)A$ was minimized.

An alternative, but rebarbative, name for this design criterion would therefore be A_A-optimality, with A-optimality recovered when $L = I$, the identity matrix.

Another special case of (10.15) arises when $s = 1$, so that A becomes the c of the previous section. If several contrasts are of interest, these can be written as the rows of the $p \times s$ matrix C, when the criterion function is $\mathrm{tr}\{C^T M^{-1}(\xi)C\}$, whence the name C-optimality.

In the notation of (10.15), the equivalence theorem states that, for the optimum design,

$$f^T(x)M^{-1}(\xi^*)AA^T M^{-1}(\xi^*)f(x) \leqslant \mathrm{tr}\{A^T M^{-1}(\xi^*)A\},$$

the generalization of the condition for c-optimality given in (10.13).

10.7 V (average variance)-optimality

A special case of c-optimality mentioned above was minimization of $f(x_0)^T M^{-1}(\xi)f(x_0)$, the variance of the predicted response at x_0. Suppose now that interest is in the average variance over a region R. Suppose further that averaging is with respect to a probability distribution μ on R. Then the design should minimize

$$\int_R f^T(x)M^{-1}(\xi)f(x)\mu(\mathrm{d}x). \tag{10.16}$$

Then, if we let

$$L = \int_R f^T(x)f(x)\mu(\mathrm{d}x),$$

which is a non-negative definite matrix, it follows that (10.16) is another example of the linear optimality criterion (10.15).

The idea of V-optimality was mentioned briefly in §6.3. In practice, the importance of the criterion is often as a means of comparing designs found by other criteria. The numerical value of the design criterion (10.16) is usually found by averaging the variance over a grid in $\mathscr{X}$.

10.8 G-Optimality

G-optimality was introduced in §6.3 and used in §9.4 in the iterative construction of a D-optimum design. The definition is repeated here for completeness.

Let

$$\bar{d}(\xi) = \max_{x \in \mathscr{X}} d(x, \xi).$$

Then the design which minimizes $\bar{d}(\xi)$ is G-optimum. For continuous designs this optimum design measure ξ^* will also be D-optimum and $\bar{d}(\xi^*) = p$. However, Example 9.3 and Fig. 9.3 show that this equivalence may not hold for exact designs. For an exact D-optimum design we may have $\bar{d}(\xi_N^*) > p$.

10.9 Other criteria and further reading

The alphabetical nomenclature for design criteria was introduced by Kiefer (1959). This paper, together with the publication of the first equivalence theorem by Kiefer and Wolfowitz (1960), ushered in a decade of rapid development of optimum design theory. The relevant literature can be found from the annotated bibliography of Herzberg and Cox (1969). More recent references are given in the survey papers of Atkinson (1982a, 1988).

The present chapter has covered the criteria to be met with in the rest of this book, apart from the model-discriminating designs of Chapter 20. These are sometimes called T-optimum, to recall the relationship with testing models. The criteria described are all functions of the information matrix $M(\xi)$. If ξ_1 and ξ_2 are two designs such that $M(\xi_1) - M(\xi_2)$ is positive definite, then ξ_1 will be a better design than ξ_2 for any criterion function Ψ. If a ξ_1 can be found for which this difference is at least non-negative definite for all ξ_2 and positive definite for some ξ_2, then ξ_1 is a globally optimum design. In most situations this is too strong a requirement to be useful, although it holds for some designs for qualitative factors, such as Latin squares. An introduction is given by Wynn (1984) in a paper which reviews Kiefer's work on optimum experimental design and includes a list of his publications.

11
D-optimum designs

11.1 Properties of D-optimum designs

In this section we list a variety of results for D- and G-optimum designs. In §11.2 an illustration is given of the use of the variance function in the iterative construction of a D-optimum design. The next section returns to the example of the desorption of carbon monoxide with which the book opened. A comparison is made of the design generating the data of Table 1.1 with several of the D- and D_s-optimum designs derived in later chapters. The last two sections of the chapter list D-optimum designs which might be useful in practice, particularly for second-order models. But, to begin, we consider general properties of D-optimum designs.

1. The D-optimum design ξ^* maximizes $|M(\xi)|$ or, equivalently, minimizes $|M^{-1}(\xi)|$.

2. The D-efficiency of an arbitrary design ξ is defined as

$$D_{\text{eff}} = \left\{ \frac{|M(\xi)|}{|M(\xi^*)|} \right\}^{1/p}. \tag{11.1}$$

Taking the ratio of the determinants in (11.1) to the $(1/p)$th power results in an efficiency measure which is proportional to design size, irrespective of the dimension of the model. So two replicates of a design measure for which $D_{\text{eff}} = 0.5$ would be as efficient as one replicate of the optimum measure.

3. A generalized G-optimum design over the region $\mathcal{R}$ is one for which

$$\max_{x \in \mathcal{R}} w(x)d(x, \xi^*) = \min_{\xi} \max_{x \in \mathcal{R}} w(x)d(x, \xi).$$

The equivalence theorem is given in Table 10.1. Usually $\mathcal{R}$ is taken as the design region $\mathcal{X}$ and $w(x) = 1$, when the equivalence of D- and G-optimum designs results. Then, with

$$\bar{d}(\xi) = \max_{x \in \mathcal{X}} d(x, \xi),$$

the G-efficiency of a design ξ is given by

$$G_{\text{eff}} = \frac{\bar{d}(\xi^*)}{\bar{d}(\xi)} = \frac{p}{\bar{d}(\xi)}. \tag{11.2}$$

4. The *D*-optimum design need not be unique. If ξ_1^* and ξ_2^* are *D*-optimum designs, the design

$$\xi^* = c\xi_1^* + (1-c)\xi_2^* \qquad (0 \leqslant c \leqslant 1)$$

is also *D*-optimum.

5. The *D*-optimality criterion is model dependent. However, the design is invariant to non-degenerate linear transformation of the model. Thus a design *D*-optimum for the model $\eta = \beta^{\mathrm{T}} f(x)$ is also *D*-optimum for the model $\eta = \gamma^{\mathrm{T}} g(x)$, if $g(x) = Af(x)$ and $|A| \neq 0$. Here β and γ are both $p \times 1$ vectors of unknown parameters.

6. The number of support points of the design is n. We have already discussed the result that there exists a *D*-optimum ξ^* with $p \leqslant n \leqslant p(p+1)/2$, although, from point 4 above, there may also be optimum designs with the same information matrix but with n greater than this limit.

7. The determinants of the information matrices of the *D*-optimum continuous design ξ^* and the *D*-optimum *N*-trial exact design ξ_N^* satisfy

$$1 \leqslant \frac{|M(\xi^*)|}{|M(\xi_N^*)|} \leqslant \frac{N^p}{N(N-1)\ldots(N+1-p)}.$$

8. If the design ξ^* is *D*-optimum with the number of support points $n = p$, then $\xi_i = 1/p$, $(i = 1, \ldots, n)$. This design will clearly be a *D*-optimum exact design for $N = p$. For these designs

$$\mathrm{cov}(\hat{Y}_i, \hat{Y}_j) \propto d(x_i, x_j) = f^{\mathrm{T}}(x_i) M^{-1}(\xi^*) f(x_j) = 0$$

$$(i, j = 1, 2, \ldots, n; \ i \neq j). \qquad (11.3)$$

This result, which also holds for non-optimum ξ with $\xi_i = 1/p$ and $n = p$, is of particular use in the construction of mixture designs with blocking (§14.1).

Other results on *D*-optimum designs can be found in the references cited at the end of this chapter. It is important to note, from the practical point of view, that *D*-optimum designs often perform well according to other criteria. The comparisons made by Donev and Atkinson (1988) for response surface designs are one example.

11.2 The sequential construction of *D*-optimum designs

In this section we give an example of the sequential construction of a *D*-optimum continuous design. We use the special case of the first-order algorithm of §9.4 which sequentially adds a trial at the point where $d(x, \xi_N)$ is a maximum. In this way a near-optimum design is constructed. However, there is no intention that the design should be performed in this manner, one trial at a time. The purpose is to find the optimum design measure ξ^*.

The algorithm can be described in a way which is helpful for the algorithms for exact designs of Chapter 15. Let the determinant of the information matrix after N trials be

$$\Delta_N = |F^T F|,$$

Then addition of one further trial at x yields the determinant

$$\Delta_{N+1} = |F^T F + f(x) f^T(x)|.$$

But (e.g. Rao 1973, p. 32),

$$\Delta_{N+1} = |F^T F| \{1 + f^T(x)(F^T F)^{-1} f(x)\}. \qquad (11.4)$$

Therefore

$$\Delta_{N+1} = \Delta_N \left\{ 1 + \frac{d(x, \xi_N)}{N} \right\}, \qquad (11.5)$$

so that the addition to the design of a trial where $d(x, \xi_N)$ is a maximum will result in the largest possible increase in Δ_N.

Example 11.1 Cubic regression through the origin

As an example of the use of (11.5) in constructing a design we take

$$\eta(x) = \beta_1 x + \beta_2 x^2 + \beta_3 x^3$$

with $\mathscr{X} = [0, 1]$. This model is chosen because it provides a simple illustration of the procedure for constructing designs in a situation for which we have not yet discussed D-optimality. Such a third-order polynomial is unlikely to be required in practice; transformations of response or explanatory variable are likely to be preferable.

The starting point for the algorithm is not crucial. We take the symmetrical three-point design

$$\xi_3 = \begin{Bmatrix} 0.1 & 0.5 & 0.9 \\ 0.33 & 0.33 & 0.33 \end{Bmatrix}. \qquad (11.6)$$

Figure 11.1(a) shows the resulting plot of $d(x, \xi_3)$. As with Fig. 10.1 for quadratic regression through the origin, the variance is zero at the origin. The maximum value of $d(x, \xi_3)$ is 18.45 at $x = 1$, reflecting the fact that the design does not span the design region.

When a trial at $x = 1$ is added to the initial design (11.6), the plot of $d(x, \xi_4)$ is as shown in Fig. 11.1(b). Comparison with Fig. 11.1(a) shows that the variance at $x = 1$ has been appreciably reduced by addition of the extra trial. The two local maxima in the curve are now of about equal importance. The maximum value of 5.45 is at $x = 0.75$. If this point is added to the design, the resulting five-point design gives rise to the variance curve of Fig. 11.1(c). The maximum variance is now at $x = 0.25$. The six-point design including this trial gives the plot of $d(x, \xi_6)$ of Fig. 11.1(d). As with Fig. 11.1(a), the maximum value is at $x = 1$, which would be the next trial to be added to the design.

The process can be continued. Table 11.1 shows the construction of the design for up to 12 trials. The search over $\mathcal{X}$ is in steps of size 0.05. The algorithm quickly settles down to the addition, in turn, of trials at 1, 0.7, and 0.25. The value of $\bar{d}(\xi_N)$ decreases steadily, but not monotonically, towards 3 and is, for example, 3.102 for $N = 50$.

Two general features are of interest: one is the structure of the design and the other is its efficiency. The structure can be seen in Fig. 11.2, a histogram of the values of x obtained up to $N = 50$. The design is evolving towards equal numbers of trials at three values around 0.2, 0.7, and 1. The starting values for the design, marked by circles, are clearly poor. There are three possibilities for finding the D-optimum design more precisely, all of which are discussed in Chapter 9.

1. Delete the poor starting design, and either start again with a better approximation to the optimum design or continue from the remainder of the design of Fig. 11.2. The deletion, as well as the addition, of trials is important in some algorithms for exact designs, such as DETMAX, described in Chapter 15.

2. Use a numerical method to find an optimum continuous design with a starting point for the algorithm suggested by Fig. 11.2.

3. Analytical optimization.

In this case we explore the third method. It is clear from the results of Fig. 11.1 that the optimum continuous design will consist of equal weight at three values of x, one of which will be unity. This structure is the same as that for the D-optimum designs discussed in earlier chapters for other polynomial models in one variable. Here $|M(\xi)|$ is a function of only two variables, and techniques similar to those of §9.5 can be used to find the optimum design. Elementary, but lengthy, algebra yields the design

$$\xi^* = \left\{ \begin{matrix} (5 - \sqrt{5})/10 & (5 + \sqrt{5})/10 & 1 \\ 1/3 & 1/3 & 1/3 \end{matrix} \right\}, \tag{11.7}$$

i.e. equal weight at $x = 0.2764, 0.7236$, and 1. A plot of $d(x, \xi^*)$ shows that this is the D-optimum design, with $\bar{d}(\xi^*) = 3$ at these design points. If the

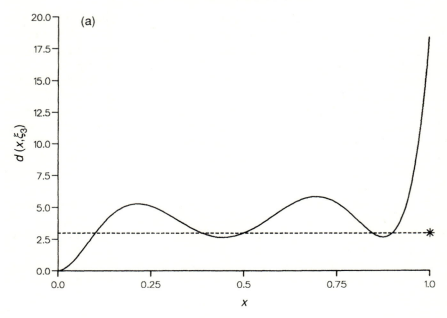

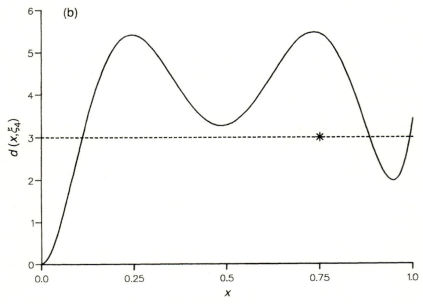

Fig. 11.1 (a and b).

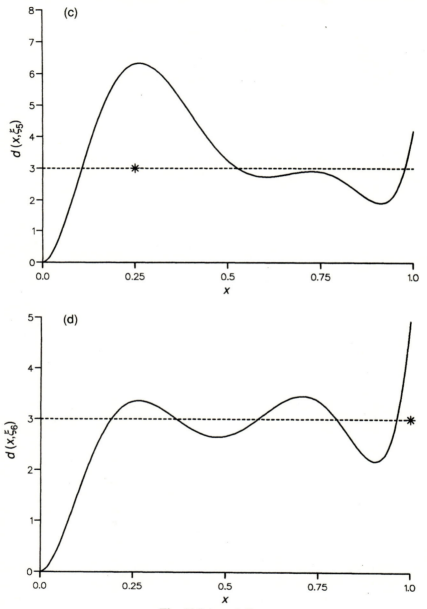

Fig. 11.1 (c and d).

Example 11.1: cubic regression through the origin. Sequential construction of the *D*-optimum design. Plots of $d(x, \xi_N)$: (a) $N=3$; (b) $N=4$; (c) $N=5$; (d) $N=6$. The asterisk is at $\bar{d}(\xi_N)$ and gives the value of x_{N+1}, the next point to be added to the design.

Table 11.1. Example 11.1: sequential construction of a *D*-optimum design for a cubic model through the origin

N	x_{N+1}	$\bar{d}(\xi_N)$	G_{eff}	D_{eff}
3	1.0	18.45	0.163	0.470
4	0.75	5.45	0.550	0.905
5	0.25	6.30	0.476	0.904
6	1.0	4.93	0.609	0.949
7	0.7	3.94	0.761	0.966
8	0.25	4.43	0.677	0.961
9	1.0	4.01	0.748	0.974
10	0.7	3.58	0.837	0.979
11	0.25	3.92	0.766	0.976
12	1.0	3.68	0.814	0.982

The initial design has trials at 0.1, 0.5, and 0.9.

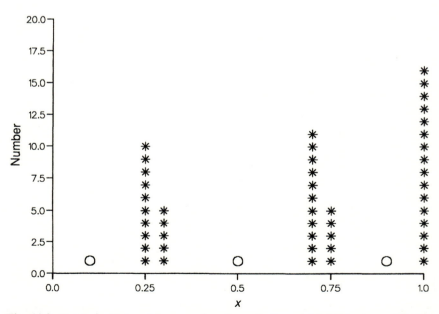

Fig. 11.2. Example 11.1: cubic regression through the origin. Histogram of design points generated by sequential construction up to $N = 50$: ○ initial three-trial design.

assumption that the design was of this form were incorrect, the plot would have revealed this through the existence of a value of $d(x, \xi) > 3$.

The *D*-efficiency of the sequentially constructed design, as defined in (11.1), is plotted in Fig. 11.3. Although the efficiency of the initial design (11.6) is only 0.470, the efficiency rises very rapidly towards unity. After only one design point has been added using the algorithm, the efficiency has risen to 0.905. An interesting feature is that progress towards the optimum is not monotonic. This feature is displayed more clearly in the plot of *G*-efficiency (Fig. 11.4). As

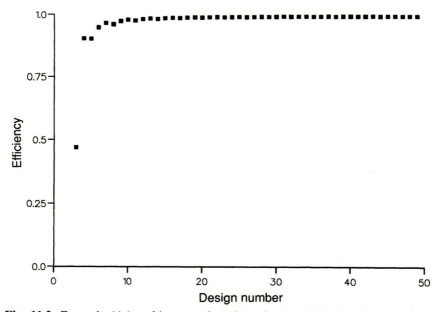

Fig. 11.3. Example 11.1: cubic regression through the origin. *D*-efficiency of the sequentially constructed design.

the plots of variance in Fig. 11.1 indicate, these efficiency values are lower than those for *D*-efficiency. They also exhibit an interesting pattern of groups of three increasing efficiency values. The highest of each group corresponds to the balanced design with nearly equal weight at the three support points. Optimum addition of one further trial causes the design to be slightly unbalanced in this respect, and leads to a decrease in *G*-efficiency. As the weight of the added trial is $1/N$, the resulting non-monotone effect decreases as N increases. A last comment on design efficiency is that the neighbourhood of the *D*-optimum design is usually fairly flat when considered as a function of ξ,

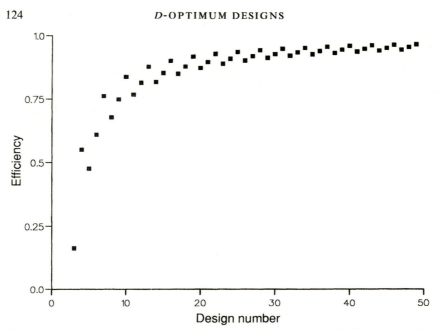

Fig. 11.4. Example 11.1: cubic regression through the origin. *G*-efficiency of the sequentially constructed design.

so that many seemingly rather different designs may have similar *D*-efficiencies.

11.3 An example of design efficiency: the desorption of carbon monoxide

(Example 1.1 continued)

The design of Table 1.1 for studying the desorption of carbon monoxide is typical of many in the scientific and technical literature: the levels of the factors and the number of replicates at each design point seem to have been chosen with no very clear objective in mind. As the examples of optimum designs in this book have shown, precisely defined objectives lead to precisely defined designs, the particular design depending upon the particular objectives. In this section the efficiency of the design of Table 1.1 is calculated for a number of plausible objectives.

The models considered for these data in Chapter 8 included first- and second-order polynomials, either through the origin or with allowance for a non-zero intercept. The *D*-optimum designs for all of these models require at most three design points: 0, 0.5, and 1. The D_s-optimum designs for checking the necessity of the two second-order models add one extra design point, $\sqrt{2} - 1 = 0.414$, and introduce unequal weighting on the design points. Even so, the design of Table 1.1, with six design points, can be expected to be inefficient for these purposes. That this is so is shown by the results of Table 11.2. Three of the D-efficiencies are below 50 per cent and only one is much above.

Table 11.2. Example 1.1: efficiency of design* for measuring the desorption of carbon monoxide

Model $\eta(x)$	Optimality criterion	Weight at design points				Efficiency (%)
		0	$\sqrt{2}-1$	0.5	1	
$\beta_0+\beta_1 x$	D	1/2	—	—	1/2	69.5
$\beta_0+\beta_1 x+\beta_2 x^2$	D	1/3	—	1/3	1/3	81.7
$\beta_0+\beta_1 x+\beta_2 x^2$	D_s for β_2	1/4	—	1/2	1/4	47.4
$\beta_1 x$	D	—	—	—	1	43.7
$\beta_1 x+\beta_2 x^2$	D	—	—	1/2	1/2	62.4
$\beta_1 x+\beta_2 x^2$	D_s for β_2	—	$\sqrt{2}/2$	—	$1-\sqrt{2}/2$	47.2

*The design region is scaled to be $\mathscr{X}=[0, 1]$. The design of Table 1.1 is then

$$\xi_{22} = \begin{Bmatrix} 0.02 & 0.1 & 0.2 & 0.5 & 0.84 & 1.0 \\ 2/22 & 2/22 & 3/22 & 5/22 & 6/22 & 4/22 \end{Bmatrix}.$$

The design is more effcient for models which are allowed to have a non-zero intercept. However, since carbon monoxide is not desorbed in the absence of the catalyst, it would seem to have been safe to have designed the experiment on the assumption that $\beta_0=0$. Then, for any of the criteria in the bottom half of the table, the optimum design would concentrate on trials at unity and near zero. With its greater spread of design points, the actual design achieves efficiencies of 40–60 per cent for these models. Thus something approaching half the experimental effort is wasted. As we have seen, a first-order model fits these data very well. It is therefore unnecessary to design for any more complicated models than those listed in the table. However, there does remain the question as to how efficient a design can be found for all, or several, of these criteria. In particular, can designs be found which are efficient both for estimating the parameters of a first-order model and for checking the fit of the model? We discuss this topic further in Chapter 21 on composite design criteria.

11.4 Polynomial regression in one variable

In the remaining two sections of this chapter we present designs for polynomial models. In this section the model is a dth-order polynomial in one factor. In the next section it is a second-order polynomial in m factors.

The model is

$$E(Y) = \beta_0 + \sum_{j=1}^{d} \beta_j x^j \qquad (11.8)$$

with $\mathcal{X} = [-1, 1]$. We begin with D-optimum designs and then consider D_s-optimum designs for the highest-order term in (11.8).

The D-optimum continuous designs for $d=1$ and $d=2$ have appeared several times. For $d=1$ half the trials are at $x=1$ and the other half are at $x=-1$. For the quadratic model, $d=2$ and a third of the trials are at $x=-1$, 0, and 1. In general, $p=d+1$ and the design puts mass $1/p$ at p distinct design points. Guest (1958) shows that the location of these points depends upon the derivative of the Legendre polynomial $P_d(x)$. This set of orthogonal polynomials is defined by the recurrence

$$(d+1)P_{d+1}(x) = (2d+1)xP_d(x) - dP_{d-1}(x) \qquad (11.9)$$

with $P_0(x) = 1$ and $P_1(x) = x$ (see, for example, Abramowitz and Stegun 1965, p. 342). From (11.9)

$$P_2(x) = \frac{3x^2 - 1}{2}$$

and

$$P_3(x) = \frac{5x^3 - 3x}{2}.$$

Guest (1958) shows that the points of support of the D-optimum design for the dth-order polynomial are at ± 1 and the roots of the equation

$$P_d'(x) = 0.$$

Equivalently (Fedorov 1972, p. 89), the design points are the roots of the equation

$$(1 - x^2)P_d'(x) = 0.$$

For example, when $d=3$, the design points are at ± 1 and those values for which

$$P_3'(x) = \frac{15x^2 - 3}{2} = 0,$$

i.e. $x = \pm 1/\sqrt{5}$. Table 11.3 gives analytical and numerical expressions for the optimum x values up to the sixth-order polynomial. The designs up to this order were first found by Smith (1918) in a remarkable paper; the design criterion was what is now called G-optimality. Her description of the criterion is as follows: 'in other words the curve of standard deviation with the lowest possible maximum value within the working range of observations is what we shall attempt to find'. In a paper of 85 pages she found designs not only for constant error standard deviation, but also for deviations of the asymmetrical form $\sigma(1+ax)\,(0\leqslant a<1)$ and of the symmetrical form $\sigma(1+ax^2)\,(a>-1)$. However, we continue to find designs for constant standard deviation. D-optimum designs for general non-constant variance are described in §22.5.

Table 11.3. Polynomial regression in one variable: points of support of D-optimum designs for dth-order polynomials

d	x_1	x_2	x_3	x_4	x_5	x_6	x_7
2	-1			0			1
3	-1		$-a_3$		a_3		1
4	-1		$-a_4$	0	a_4		1
5	-1	$-a_5$	$-b_5$		b_5	a_5	1
6	-1	$-a_6$	$-b_6$	0	b_6	a_6	1

$a_3=1/\sqrt{5}=0.4472$
$a_4=\sqrt{(3/7)}=0.6547$
$a_5=\sqrt{\{(7+2\sqrt{7})/21\}}=0.7651$
$b_5=\sqrt{\{(7-2\sqrt{7})/21\}}=0.2852$
$a_6=\sqrt{\{(15+2\sqrt{15})/33\}}=0.8302$
$b_6=\sqrt{\{(15-2\sqrt{15})/33\}}=0.4688$

Table 11.3 exhibits the obvious feature that the optimum design depends upon the order of the polynomial model. In §21.2 designs are found, using a composite design criterion, which are simultaneously as efficient as possible for all models up to the sixth order. The results are summarized in Tables 21.1 and 21.2. We conclude the present section with the special case of D_s-optimum designs for the highest-order term in the model (11.8).

For the quadratic polynomial, i.e. $d=2$, the D_s-optimum design for β_2 when $\mathcal{X}=[-1,1]$ puts half the trials at $x=0$, with a quarter each at $x=-1$ and $x=+1$. The extension to precise estimation of β_d in the dth-order polynomial (Kiefer and Wolfowitz 1959) depends on Chebyshev polynomials. The design again has $d+1$ points of support, but now with

$$x_j=-\cos\left(\frac{j\pi}{d}\right) \qquad (0\leqslant j\leqslant d)$$

The D_s-optimum design weight w^* is spread equally over the $d-1$ points in the interior of the region with the same weight divided equally between $x = \pm 1$, i.e.

$$w^*(-1) = w^*(1) = \frac{1}{2d}$$

$$w^*\left\{\cos\left(\frac{j\pi}{d}\right)\right\} = \frac{1}{d} \qquad (1 \leqslant j \leqslant d-1),$$

which agrees with the 1/4, 1/2, 1/4 weighting when $d = 2$. If exact designs are required, the numerical methods of Chapter 15 have to be employed.

11.5 Second-order models

The second-order polynomial in m factors is

$$E(Y) = \beta_0 + \sum_{j=1}^{m} \beta_j x_j + \sum_{j=1}^{m-1} \sum_{k=j+1}^{m} \beta_{jk} x_j x_k + \sum_{j=1}^{m} \beta_{jj} x_j^2.$$

Continuous D-optimum designs for this model over the sphere, cube, and simplex are given by Farrell et al. (1967), who also give designs for higher-order polynomials. In this section we first consider designs when $\mathcal{X}$ is a sphere and then when it is a cube. In both cases the description of the optimum continuous design is augmented by a table of small exact designs. Designs over the simplex are the subject of Chapter 12 on mixture experiments.

D-optimum continuous designs over the sphere have a very simple structure. A measure $2/\{(m+1)(m+2)\}$ is put at the origin, i.e. the centre point of the design. The rest of the design weight is uniformly spread over the sphere of radius $\sqrt{m}$ which forms the boundary of $\mathcal{X}$. Table 11.4 gives the values of $|M(\xi^*)|$ for these optimum designs for small m, together with the values of $\bar{d}(\xi^*)$, which equal p, and the values of $d_{\text{ave}}(x, \xi^*)$ found, for computational convenience, by averaging over the points of the 5^m factorial with vertices ± 1. Although this averaging excludes part of $\mathcal{X}$, it does provide an interesting basis for the comparison of designs.

Exact designs approximating the continuous designs are found by the addition of centre points and star points, with axial co-ordinate $\sqrt{m}$, to the points of the 2^m factorial. Table 11.4 gives nine such designs, for several of which the D-efficiency is 98 per cent or better. The addition of several centre points to the designs, which is often recommended to provide an estimate of σ^2, causes a decrease in D-efficiency. For example, for $m = 3$ the optimum weight at the centre is $2/\{(m+1)(m+2)\} = 1/10$, so that the addition of one or two centre points to the 2^3 factorial with star points provides a good exact design. However, increasing the number of centre points does initially have the desirable effect of reducing the average and maximum values of $d(x, \xi)$. In

Table 11.4. Second-order polynomial in m factors: spherical experimental region

(a) *Continuous D-optimum designs*

m	p	$|M(\xi^*)|$	d_{ave}	d_{max}
2	6	2.616×10^{-2}	4.40	6.0
3	10	2.519×10^{-7}	7.14	10.0
4	15	7.504×10^{-15}	10.71	15.0
5	21	4.440×10^{-25}	15.17	21.0

(b) *Central composite exact designs*

m	p	N	N_0	D-efficiency (%)	d_{ave}	d_{max}
2	6	9	1	98.6	5.517	9.000
2	6	11	3	96.9	4.345	6.875
2	6	13	5	89.3	4.569	8.125
3	10	15	1	99.2	8.159	15.00
3	10	17	3	97.7	6.542	10.52
3	10	19	5	91.9	6.707	11.76
4	15	25	1	99.2	12.37	25.00
4	15	27	3	98.9	10.63	15.75
4	15	29	5	95.2	10.83	16.92

The designs consist of a 2^m factorial with $2m$ star points and N_0 centre points. All values of d_{ave}, as well as d_{max} for the exact designs, are calculated over the points of the 5^m factorial.

interpreting the results of Table 11.4 it needs to be kept in mind that the values of d_{ave} and the maximum variance d_{max} are calculated only at the points of the 5^m factorial. In particular, the values of d_{max} for the number of centre points $N_0 = 1$ are an underestimate of $\bar{d}(\xi)$, which equals p only for the optimum continuous design.

The situation for cubic design regions is slightly more complicated. Farrell *et al.* (1967) show that the optimum continuous design is supported on subsets of the points of the 3^m factorial, with the members of each subset having the same number of non-zero co-ordinates. Only three subsets are required, over each of which a specified design weight is uniformly distributed. To identify the three sets of points let, for example, $j_1 = 0$ be the centre point and $j_3 = m$ be the set of points of the 2^m factorial, i.e. with all m co-ordinates equal to ± 1, which are the corner points of the 3^m factorial. The D-optimum continuous designs then have support on subsets with

$$0 \leqslant j_1 \leqslant m-2 \qquad j_2 = m-1 \qquad\qquad j_3 = m \quad (2 \leqslant m \leqslant 5)$$

$$0 \leqslant j_1 \leqslant m-3 \qquad j_2 = m-2 \text{ or } m-1 \qquad j_3 = m \qquad (m \geqslant 6). \quad (11.10)$$

Of the designs satisfying (11.10), those with support $(0, m-1, m)$ require fewest distinct design points. These are the centre point, the midpoints of edges, and the corner points of the 3^m factorial, respectively. This family was studied by Kôno (1962). The weights for the D-optimum continuous designs for $m \leqslant 5$ are given in Table 11.5.

Table 11.5. Second-order polynomial in m factors: cubic experimental region. Weights for optimum continuous D-optimum designs supported on points of the 3^m factorial with 0, $m-1$, and m non-zero co-ordinates

Number of factors m	Design weights		
	w_0	w_{m-1}	w_m
2	0.096	0.321	0.583
3	0.066	0.424	0.510
4	0.047	0.502	0.451
5	0.036	0.562	0.402

It is interesting to note that central composite designs, which belong to the family $(0, 1, m)$, cannot provide the support for an optimum continuous design. However, they may provide adequate integer approximations to the continuous D-optimum designs. This topic is explored further in §21.3 where Table 11.5 is extended to include D_s-optimum designs for subsets of the parameters in the quadratic model and where exact central composite designs are assessed. In the remainder of this section we present exact designs which do not have the symmetry of the composite designs.

Since the D-optimum continuous designs have support on the points of the 3^m factorial, it is reasonable to expect that good exact designs for small N can be found by searching over the points of the 3^m factorial. Table 11.6 lists designs for second-order models for $m \leqslant 5$ found using the KL exchange algorithm of Chapter 15 for this search. The design points are labelled in standard order for the 3^m factorial. Examples of this notation are given in

Table 11.7. The design points in Table 11.6 are listed in order of decreasing variance. In some cases the optimum designs exhibit a sequential property: for example, the design for $m=2$, $N=7$ is found by adding the point 5 to the design for $m=2$, $N=6$.

Table 11.6. Second-order polynomial in m factors: cubic experimental region. Exact D-optimum designs supported on points of the 3^m factorial

m	N	Design points
2	6	1 3 4 7 8 9
2	7	($m=2$, $N=6$) 5
2	8	($m=2$, $N=7$) 6
3	10	27 19 3 8 16 21 1 23 15 11
3	14	7 3 19 21 9 25 1 5 15 11 27 13 17 23
3	15	($m=3$, $N=10$) 9 25 4 7 2
3	16	($m=3$, $N=15$) 6
3	18	($m=3$, $N=16$) 21 19
3	20	($m=3$, $N=18$) 18 26
4	15	54 7 38 78 22 6 57 73 58 79 21 26 1 63 18
4	18	10 66 9 41 55 25 21 79 30 34 46 81 60 4 62 74 2 27
4	24	72 36 54 3 55 21 25 73 1 18 40 38 61 59 75 80 7 19 57 6 79 78 8 27
4	25	5 75 69 49 21 57 1 38 79 62 55 9 25 54 34 19 3 73 70 26 63 18 7 81 58
4	27	($m=4$, $N=25$) 24 10
5	21	189 21 81 1 46 217 57 241 169 61 225 25 9 165 64 181 237 157 204 125 77
5	26	55 157 126 70 223 235 209 14 19 187 219 171 63 7 27 3 102 243 163 137 75 52 193 186 80 174
5	27	169 114 216 44 229 218 21 237 73 61 165 81 57 4 225 25 9 185 181 241 91 108 190 2 180 155 55

Designs consist of the points of the 3^m factorial labelled in standard order and listed in order of decreasing variance (see Table 11.7 for standard order notation).

Properties of the designs given in Table 11.8 include the D- and G-efficiencies relative to the continuous designs of Table 11.5. For fixed m, and therefore fixed number of parameters p, the D-efficiency is smallest for $N=p$. The addition of one or two trials causes an appreciable increase in the efficiency of the design, in addition to the reduced variance of parameter estimates coming from a larger design. This effect decreases as m increases. The G-efficiencies and the values of d_{ave} were calculated over a 5^m grid, as in

Table 11.7. Examples of the standard order notation used in Table 11.6 to describe the points of the 3^m factorial

m	Point number	x_1	x_2	x_3	x_4	x_5
2	1	-1	-1			
	2	0	-1			
	3	1	-1			
	7	-1	1			
3	19	-1	-1	1		
	20	0	-1	1		
4	26	0	1	1	-1	
	27	1	1	1	-1	
	28	-1	-1	-1	0	
	80	0	1	1	1	
5	229	-1	0	0	1	1

Table 11.4. The general behaviour of G-efficiency is similar to that of D-efficiency; for instance, moving from $N=p$ to $N=p+1$ produces a large increase. However, as comparison of Figs 11.3 and 11.4 showed, the behaviour of G-efficiency is more volatile than that of D-efficiency, although the trend to increasing efficiency with N is evident. Small values of the average variance d_{ave} are desirable and these behave much like the reciprocal of G-efficiency, yielding better values as N increases for fixed m.

The designs of Tables 11.4 and 11.6 should meet most practical situations where a second-order polynomial model is to be fitted. Methods of dividing the design into blocks are given in Chapter 14. Chapter 16 describes how to proceed if the design region is restricted and so does not have the regular shape of a sphere or a cube.

11.6 Further reading

The study of D-optimality has been central to work on optimum experimental design since the beginning (e.g. Kiefer 1959). An appreciable part of the material in the books by Fedorov (1972), Silvey (1980), and Pazman (1986) likewise stresses D-optimality. Farrell *et al.* (1967), in addition to the results quoted in this chapter, give a summary of earlier work on D-optimality. This includes Kiefer and Wolfowitz (1959) and Kiefer (1961) which likewise concentrate on results for regression models, including extensions to D_s-optimality.

Table 11.8. Second-order polynomial in m factors: cubic experimental region. Properties of the exact D-optimum designs of Table 11.6

m	N	p	D_{eff}	G_{eff}	d_{ave}
2	6	6	0.8849	0.3636	8.78
2	7	6	0.9454	0.6122	5.74
2	8	6	0.9572	0.6000	5.38
2	9	6	0.9740	0.8276	4.89
3	10	10	0.8631	0.2904	11.22
3	14	10	0.9759	0.8929	7.82
3	15	10	0.9684	0.7752	8.85
3	16	10	0.9660	0.7418	8.81
3	18	10	0.9717	0.6817	9.22
3	20	10	0.9779	0.8258	8.93
4	15	15	0.8700	0.4522	19.28
4	18	15	0.9311	0.5114	14.50
4	24	15	0.9670	0.6640	14.16
4	25	15	0.9773	0.6890	14.07
4	27	15	0.9815	0.6983	13.89
5	21	21	0.9055	0.4471	26.62
5	26	21	0.9519	0.6705	19.86
5	27	21	0.9539	0.6494	19.71

Silvey (1980) compares first-order algorithms of the sort exemplified in §11.2. If, perhaps as a result of such a sequential construction, the support of the optimum design is clear, the weights of the continuous design can be found by numerical optimization. Alternatively, a special algorithm, such as that of Silvey *et al.* (1978) can be used. Further fractions of the 3^m factorial extending the results of Table 11.6 can be found using the BLKL algorithm of Appendix A. Methods for the augmentation of computer search by the use of orthogonal fractions of 2^m and 3^m factorials are described by Mitchell (1974*b*), Mitchell and Bayne (1978), and Pesotchinsky (1975).

12
Mixture experiments

12.1 Introduction

This chapter is concerned with the design of experiments when there is a constraint on q of the factors

$$\sum_{i=1}^{q} x_i = 1 \qquad (x_i \geqslant 0). \qquad (12.1)$$

We shall chiefly consider the case when the remaining factors, if any, do not vary during the experiment. Then (12.1) defines the q components of a mixture. In mixture experiments the response will depend only on the proportions of the components in the mixture, for example an alloy, but not on the total amount. Experiments of this kind occur frequently in such areas as chemistry, medicine, and agriculture. An extensive survey of mixture experiments is given by Cornell (1990).

The general theory of the optimum design of experiments applies to mixture experiments. However, constraint (12.1) introduces some special features. In particular, changes in the values of one of the factors will lead to changes in the value of at least one of the other factors. The design region becomes a $(q-1)$-dimensional regular simplex and the ordinary polynomial models are inappropriate.

The choice of model for a mixture experiment is not always straightforward. Some of the many models in the literature are described in the next section, together with some classical designs. Once an appropriate model has been chosen, the optimum design can be calculated in the standard way. In Section 12.3 some consideration is given to the situation where the experiment is to be performed in a constrained part of the simplex. We also show how mixture experiments in which the response depends on both the proportions of the q components and the amount of the mixture can be regarded as a $(q+1)$-component mixture problem. In the last section we consider briefly an experiment in which there are both mixture variables and a qualitative factor. Blocking of mixture experiments is discussed in Chapter 14. Examples of the design of mixture experiments for arbitrarily restricted regions are given in Chapter 16.

12.2 Models and designs for mixture experiments

The canonical polynomials of Scheffé (1958) have been widely applied because of the flexible family of models they provide. They are obtained by reparameterization of standard polynomials allowing for the relationship between the factors following from (12.1). The first-order Scheffé polynomial is

$$E(Y) = \sum_{i=1}^{q} \beta_i x_i, \tag{12.2}$$

which does not explicitly contain a constant term, while the second-order model becomes

$$E(Y) = \sum_{i=1}^{q} \beta_i x_i + \sum_{i=1}^{q-1} \sum_{j=i+1}^{q} \beta_{ij} x_i x_j. \tag{12.3}$$

In (12.3) the effect of the constraint is to render redundant the pure quadratic terms. For higher-order models the reparameterization does not lead to such simple expressions. For example, a symmetrical way of writing the third-order model is

$$E(Y) = \sum_{i=1}^{q} \beta_i x_i + \sum_{i=1}^{q-1} \sum_{j=i+1}^{q} \beta_{ij} x_i x_j + \sum_{i=1}^{q-1} \sum_{j=i+1}^{q} \gamma_{ij} x_i x_j (x_i - x_j)$$

$$+ \sum_{i=1}^{q-2} \sum_{j=i+1}^{q-1} \sum_{k=j+1}^{q} \beta_{ijk} x_i x_j x_k. \tag{12.4}$$

For estimation of the parameters in these canonical polynomials, Scheffé (1958) proposed 'simplex lattice' designs. For the polynomial of degree d these designs include all possible combinations of trials in which each component takes values $0, 1/d, 2/d, \ldots, 1$. Such designs are saturated, i.e. the number of trials is equal to the number of parameters in the model. This simplifies the expression for $M(\xi)$ and leads to simple formulae for the estimation of the parameters. Like other saturated designs, simplex lattice designs suffer from the disadvantage that they provide no information on lack of fit; additional trials have to be added to check the adequacy of the model. A specific shortcoming of the designs is that they contain a high proportion of trials for which at least one component takes the value zero. If the behaviour of the complete mixture is markedly different from the behaviour of simpler systems lacking one or more components, the information provided by the simplex lattice designs will be seriously inadequate.

The D-optimality of the simplex lattice designs for the first- and second-order canonical polynomials was established by Kiefer (1961). As an example Fig. 12.1 shows the six-trial D-optimum design for the second-order polynomial (12.3) when $q = 3$. This is also the optimum continuous design, the measure putting weight 1/6 at the vertices and middles of the edges of the triangle representing the design region. However, this optimality result does not extend to higher-order models.

Figure 12.2 shows the third-order simplex lattice design which has support at the vertices of the design region, at the centre, and on the edges at points corresponding to all two-component mixtures in which one of the components is one-third of the mixture and the other is two-thirds. For this design the determinant of the information matrix $|M(\xi_{10})| = 0.1747 \times 10^{-21}$. The D-optimum design for (12.4) with $q = 3$ has the same structure except that the (1/3, 2/3) mixtures are replaced by those derived from the results on D-optimality for dth-order polynomials in Table 11.3. There, for the third-order polynomial over $\mathscr{X} = [-1, 1]$, the interior design points were at $x = \pm 1/\sqrt{5}$. Rescaling of these points on to [0, 1] yields design points at $(1 \pm 1/\sqrt{5})/2$, which are the proportions for the optimum two-component mixtures, namely (0.2764, 0.7246). These design points are shown in Fig. 12.2 as open circles. For the resulting design $|M(\xi^*)| = 0.2875 \times 10^{-21}$, a modest improvement over the simplex lattice design when the number of parameters is taken into account.

In order to improve the interpretation of the parameters in polynomial mixture models Cox (1971) suggests reparameterization to represent the change in the response as one factor is varied, with the relative proportions of the remaining factors being held constant. Draper and John (1977) recommend the inclusion of inverse terms into Scheffé polynomials, provided that the levels of the components never reach zero. For example, their first-order model is

$$E(Y) = \sum_{i=1}^{q} \beta_i x_i + \sum_{i=1}^{q} \frac{\beta_{-i}}{x_i},$$

which allows behaviour near to lower-dimensional mixtures to be quite different from that for the q-component mixture.

If one of the components has an additive blending effect, the models proposed by Becker (1968) may be useful. Three possibilities are

$$E(Y) = \sum_{i=1}^{q} \beta_i x_i + \sum_{i=1}^{q-1} \sum_{j=i+1}^{q} \beta_{ij} \min(x_i, x_j) + \cdots + \beta_{12\ldots q} \min(x_1, x_2, \ldots, x_q)$$

$$E(Y) = \sum_{i=1}^{q} \beta_i x_i + \sum_{i=1}^{q-1} \sum_{j=i+1}^{q} \frac{\beta_{ij} x_i x_j}{x_i + x_j} + \cdots + \frac{\beta_{12\ldots q} x_1 x_2 \ldots x_q}{x_1 + \cdots + x_q}$$

$$E(Y) = \sum_{i=1}^{q} \beta_i x_i + \sum_{i=1}^{q-1} \sum_{j=i+1}^{q} \beta_{ij}(x_i x_j)^{1/2} + \cdots + \beta_{12\ldots q}(x_1 x_2 \ldots x_q)^{1/q}.$$

These models are symmetrical in the q factors and so are only appropriate if the extremum of the surface they are modelling is at or near the centre of the

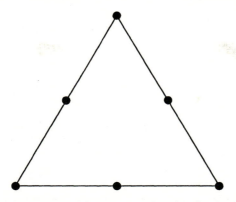

Fig. 12.1. Three-component mixture: second-order simplex lattice design.

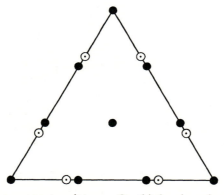

Fig. 12.2. Three-component mixture: ● third-order simplex lattice design; ○ D-optimum design for third-order model.

simplex. Becker also suggests a decentralized form of the first of these models. A further class of models (Kenworthy 1963; Becker 1969, 1970) incorporates ratios of the components.

The customary designs for several of these models are saturated. Vuchkov (1982) uses the results on saturated designs of §11.1 for design problems for canonical polynomials. In particular, the variance of the predicted response $\hat{Y}$ at the support points x_i is

$$\text{var}(\hat{Y}_i) = \sigma^2 f(x_i)^{\text{T}}(F^{\text{T}}F)^{-1}f(x_i) = \frac{\sigma^2}{n(x_i)} \qquad (i=1, \ldots, N) \qquad (12.5)$$

where, since the design is saturated, $N = p$. In (12.5), $n(x_i) > 0$ is the number of replications at the point x_i. Further, the covariance between the predicted values at the points x_i and x_j is zero. These results are helpful in §13.5 for the construction of designs when there are both mixture and qualitative factors.

12.3 Constrained mixture experiments

Often the design region for mixture experiments is restricted to part of the $(q-1)$-dimensional regular simplex. For example, gunpowder is a mixture of sulphur, carbon, and saltpetre, the interesting explosive properties of which only occur for a small part of the triangular region of possible blends. In general, the sub-area of the simplex forming the design region can have any irregular shape. The simplest constrained case is when the design region is again a $(q-1)$-dimensional regular simplex within the larger simplex. Such a situation can occur when additional constraints of the form

$$x_i \geqslant x_{i,\min} > 0$$

are imposed on some or all the factors. The design region will also again be a regular simplex if the constraints are

$$x_i \leqslant x_{i,\max} < 1$$

and the sum of the $(q-1)$ largest upper bounds is less than or equal to unity.

For constrained design regions which are a regular simplex, Kuroturi (1966) and Crosier (1984) propose the use of pseudo-components g_i $(i=1, \ldots, q)$. These are linear transformations of the original variables which allow use of the standard results for unconstrained mixture experiments. Let the columns of the $q \times q$ matrix A be the co-ordinates of the vertices of the restricted design region. Then the relationship between the co-ordinates of a mixture x in the original components and as a pseudo-component g is

$$x = Ag \qquad (12,6)$$

with inverse

$$g = A^{-1}x. \qquad (12.7)$$

The design is constructed for the pseudo-components g, with the values of the components in the design calculated from (12.6). Since the vectors x and g are $q \times 1$, the rows of the design matrices are respectively x^T and g^T. The model can be fitted either in the pseudo-components or in the original mixture components. If pseudo-components are used for fitting, the inverse transformation (12.7) is required to give estimates of the original parameters.

Example 1 Modification of an acrylonitrile powder

Garvanska *et al.* (1992) describe the development of a modified acrylonitrile powder with improved electrophysical properties. The modification to the surface of the powder used a three-component mixture of chemicals with factors the following proportions:

x_1 copper sulphate ($CuSO_4$)

x_2 sodium thiosulphate ($Na_2S_2O_3$)

x_3 glyoxal ($CHO)_2$.

Consideration of the mechanism of the reaction imposed constraints on the factor levels:

$$0.2 \leqslant x_1 \leqslant 0.8$$

$$0.2 \leqslant x_2 \leqslant 0.8$$

$$0 \leqslant x_3 \leqslant 0.6. \tag{12.8}$$

It is straightforward to verify that the resulting constrained design region is again a simplex. The matrix A (12.6) for the pseudo-components is

$$A = \begin{bmatrix} 0.8 & 0.2 & 0.2 \\ 0.2 & 0.8 & 0.2 \\ 0 & 0 & 0.6 \end{bmatrix} \tag{12.9}$$

where each column defines one of the vertices of the restricted design region.

One response of interest was the electric resistivity W per unit volume. This was measured using a Scheffé lattice design for a second-order model in the pseudo-components. The results are given in Table 12.1. To check this saturated model for lack of fit, additional trials were performed at points, not listed in Table 12.1, which were uniformly spread through the design region. Despite the very different response value for the last trial, there was no evidence of lack of fit: it was expected that the response would change rapidly in this part of the experimental region.   □

Example 12.1 illustrates the use of the pseudo-components in the design of mixture experiments when the constrained design region is again a regular simplex. However, with more general constraints of the form

$$0 \leqslant x_{i,\min} \leqslant x_i \leqslant x_{i,\max} \leqslant 1,$$

the design region will usually be an irregular simplex, so that the use of pseudo-components is not possible. As an alternative, McLean and Anderson (1966) propose extreme vertices designs which Saxena and Nigam (1973) show may contain clusters of points in the design region. Their alternative symmetric simplex designs overcome the problem of clustering, but both design strategies suffer from an excessive number of trials as the size of the problem increases. In such cases the trials generated by these procedures can be considered as a list of

Table 12.1. Example 12.1: modification of an acrylonitrile powder. Use of pseudo-components for design in a constrained region

Pseudo-components			Mixture variables			Responses			
g_1	g_2	g_3	x_1	x_2	x_3	$\log W$	$y(z_1)$	$y(z_2)$	$y(z_3)$
1	0	0	0.8	0.2	0	1.25	8	9	9
0	1	0	0.2	0.8	0	1.82	3	4.8	4
0	0	1	0.2	0.2	0.6	0.17	8	8.5	8
0.5	0.5	0	0.5	0.5	0	0.44	7.8	8.2	7.5
0.5	0	0.5	0.5	0.2	0.3	0.03	8	8.8	8
0	0.5	0.5	0.2	0.5	0.3	11.14	0.6	2	1.8

The responses $y(z_i)$ are the electromagnetic damping at the three wavelengths $z_i = 8$, 10, and 12 GHz. W is the electric resistivity per unit volume.

candidate points. A sensible number of them can then be selected by an optimum design algorithm. An example is given in Chapter 16 where we consider the design of experiments over arbitrarily constrained design regions.

12.4 The amount of a mixture

So far we have followed the convention, which applies to many mixture experiments, that the response depends only on the proportions of the mixture components. But suppose that the response also depends on the amount A of the mixture. Let this be defined to vary in the experiment between some minimum and maximum values so that

$$0 \leqslant A_{\min} \leqslant A \leqslant A_{\max}.$$

When $A_{\min}$ is strictly positive and the amount can be regarded as a qualitative factor at two levels, we can apply the results of Claringbold (1955), Vuchkov et $al.$ (1981), and Piepel and Cornell (1985), who in addition suggest models and designs for this case. The results of §13.5 on designs when both qualitative and mixture variables are present can also be applied. However, when $A_{\min} = 0$, this approach is not possible.

In the general case when the amount is a quantitative factor it is possible to define $q + 1$ pseudo-components

$$g_{ij} = \begin{cases} \dfrac{x_{ij}}{A_{\max}} & (i = 1, \ldots, q) \\[2ex] \dfrac{A_{\max} - A_j}{A_{\max}} & (i = q + 1) \end{cases} \tag{12.10}$$

which form a $(q+1)$-component mixture including the total mixture amount as well as the q actual components of the mixture (Donev 1989). In (12.10) x_{ij} is the amount of the ith component in the mixture with total mixture amount A_j, where

$$A_j = \sum_{i=1}^{q} x_{ij}.$$

The pseudo-component g_{q+1} corresponds to the total mixture amount and takes the value unity when the amount is zero, decreasing to zero when the total mixture amount increases to its maximum possible value. In this way the design and analysis of experiments for mixtures, where the total amount is a factor, can be regarded as a $(q+1)$-dimensional mixture problem, although the interpretation of the results is different.

12.5 Mixture experiments with other factors

In addition to the q mixture variables an experiment may also include one or more qualitative or quantitative factors, so that the design region will often be the product of two regions, one for the mixture variables and the other for the remaining factors. For example, with three mixture variables and one quantitative factor the design region will be a triangular prism, provided that experiments are possible at all combinations of the two sets of variables.

Example 12.1 Modification of an acrylonitrile powder (continued)

A second interesting property of the modified acrylonitrile powder of Table 12.1 was the effect of the modification on the electromagnetic damping y of the material. It was expected that this would depend on the wavelength of the electromagnetic radiation. From previous experience the effect of the wavelength was expected to be additive and representable by a vector of indicator variables e. The response was measured at three wavelengths (Table 12.2). The simplex lattice design for the three pseudo-components was repeated at each level of the indicator variables. The results are given in Table 12.1, for which the fitted model in terms of the pseudo-components is

$$\hat{y} = 0.983e_2 + 0.483e_3 + 8.178g_1 + 3.444g_2 + 7.678g_3 + 6.133g_1g_2$$
$$- 0.600g_1g_3 - 18.333g_2g_3.$$

In order to achieve identifiability of the coefficients in the model, there is no term in e_1. This corresponds to the absence of the constant term in the canonical polynomial (12.3). □

The design of Table 12.1 repeats the mixture design at each level of the

Table 12.2. Example 12.1: modification of an acrylonitrile powder. Indicator variables for the three wavelengths

Wavelength (GHz)	Indicator variables		
	e_1	e_2	e_3
8	1	0	0
10	0	1	0
12	0	0	1

qualitative factor. In the next chapter we consider the optimality of such designs and obtain exact optimum designs requiring fewer trials. But we start the chapter with similar questions about designs in which some factors are qualitative and some are quantitative.

13
Experiments with both qualitative and quantitative factors

13.1 Introduction

This chapter is concerned with the design of experiments when the response depends on both qualitative and quantitative factors. In the chemical industry the yield of a process might depend not only on the quantitative factors temperature and pressure, but also on such qualitative factors as the batch of raw material and the type of reactor. Likewise, an antibiotic might be given orally or by an injection, a qualitative factor with two levels. The composition and dosage of the antibiotic could be the quantitative factors.

There is a huge literature on optimum designs when only one kind of factor is present. For example, the papers collected in Kiefer (1985) consider both the optimality, over a wide range of criteria, of block designs and Graeco-Latin squares, and derive designs for regression over a variety of experimental regions. In contrast, very little attention has been given to designs for both classes of factor.

***Example* 13.1** Quadratic regression with a single qualitative factor (Example 5.3 continued)

In this model the response depends not only on a quadratic in a single quantitative variable x but also on a qualitative factor z at B levels. If the effect of z is purely additive, the model is

$$E(Y_i) = \sum_{j=1}^{B} \alpha_j e_{ij} + \beta_1 x_i + \beta_2 x_i^2 \qquad (i=1, \ldots, N_i; j=1, \ldots, B). \quad (13.1)$$

In the matrix notation of §5.3 this was written

$$E(Y) = W\gamma = E\alpha + F\beta \qquad (13.2)$$

where E, of dimension $N \times B$, is the matrix of indicator variables e_j taking the values 0 or 1 for the level of the qualitative factor.

The extension of the design given in §5.2 is to repeat the three-trial D-optimum design for the quadratic at each level of z. As B increases, this product design involves an appreciable number of trials for the estimation of rather few parameters. □

Example **13.2** Second-order response surface with one qualitative factor

The chief means of illustrating the structure of designs when the two kinds of factor are present will be the extension of the previous example to two quantitative factors. One application would be to model the example from the chemical industry. Suppose that the effect of the quantitative factors can be described by a quadratic response surface, the mean value depending on the value of a single qualitative factor, the levels of which represent combinations of raw material and reactor type. This formulation implies no interaction between the qualitative and quantitative factors. The model is then

$$E(Y) = \sum_{i=1}^{B} \alpha_i e_i + \beta_1 x_1 + \beta_2 x_2 + \beta_{11} x_1^2 + \beta_{22} x_2^2 + \beta_{12} x_1 x_2, \qquad (13.3)$$

when the qualitative factor is at B levels. ☐

These simple examples indicate some of the many possibilities and complications. Although the design region for the factors x will usually be the same for all levels of z, there is no reason why this should be the case: certain combinations of x_1 and x_2 might be inadmissible for some levels of the qualitative factor. The resulting restricted design region, which is similar to, but more complicated than, those of Chapter 16, presents no difficulties for the numerical calculation of exact designs. However, it is theoretically intractable. A second potential complication is that the model (13.2) might be too simple, since there could be interactions beween the two groups of factors, causing the shape of the response surface to depend on the level of the qualitative factor. However, there are some general theoretical results.

Kurotschka (1981) shows that, under certain conditions, the optimum continuous design consists of replications of the design for the quantitative factors at each level of the qualitative ones. Such designs are called product designs. This work is outlined in the next section. A disadvantage of these designs is that the number of experimental conditions needed is large compared with the number of parameters. Therefore they will often be impracticable. Accordingly, §13.3 is concerned with exact designs, particularly when the number of trials is not much greater than the number of parameters. These designs, which are often almost as efficient as the continuous product designs, exhibit several interesting properties when compared with continuous designs. One is that the numbers of trials at the levels of the qualitative factors are often not even approximately equal. A second is that, at the individual levels of the qualitative factor, designs with some structure in the quantitative factors seem to be preferred. A third, less appealing, feature is that for some values of N the addition of one extra trial can cause a rather different design to be optimum. This suggests that care may be needed in the choice of the size of the experiment. D_s-optimum designs when the qualitative factors are regarded as nuisance parameters are considered in §13.4. The case when the quantitative factors are the components of a mixture is treated in §13.5.

13.2 Continuous designs

The general model is

$$E(Y) = \eta(x, z, \gamma) \tag{13.4}$$

where x represents the quantitative factors and z the qualitative ones. The parameterization of (13.4) can be complicated, even for linear models, if there are interactions between x and z. It is convenient to follow the classification introduced by Kurotschka (1981) who distinguishes three cases:

(1) complete interaction between qualitative and quantitative factors;

(2) no interaction between qualitative and quantitative factors, although there may well be interaction within each group of factors;

(3) the intermediate case of some interaction between groups.

The model corresponding to Case 1 has parameters for the quantitative factor which are allowed to be different at each combination of the qualitative factors. The design problem then becomes that of a series of distinct design regions $\mathcal{X}_i$ $(i = 1, \ldots, B)$. The models need not all be the same. Let the model at the ith level of z have parameter vector β_i, of dimension p_i. The D-optimum continuous design for this model over $\mathcal{X}_i$ is δ_i^*. In order to find the optimum design for the whole experiment, we also need to consider the distribution of experimental effort between the levels of z. Let v be the measure which assigns mass v_i to the experiments at level i. Then the measure on the points of $\mathcal{X}_i$ is the product $v_i \times \delta_i = \xi_i$. From the Equivalence Theorem, the D-optimum design must be such that the maximum variance is the same at all design points. Therefore v_i^* must be proportional to p_i and the optimum measure is

$$\xi_i^* = \frac{p_i}{\Sigma p_i} \times \delta_i^*. \tag{13.5}$$

If the models and design regions are the same for all levels of z while the parameters remain different, the optimum design can be written in the simpler product form as

$$\xi^* = v^* \times \delta^*, \tag{13.6}$$

where $v^* = \{1/B\}$ is now a uniform measure over the levels of z. Similar conditions can also be found for A-optimality.

For Case 2, in which there is no interaction between x and z, the model has the simple form (13.2). Although the structure of the qualitative factors z may be complicated, the experiment can be regarded as one with a single qualitative factor acting on B levels which is formed from all possible combinations of the l qualitative factors. For example, with two qualitative factors we can let $E_\alpha = E_1\alpha_1 + E_2\alpha_2$, with the identifiability constraint that one of the elements of either α be set to zero. This form could represent all $l_1 \times l_2$ conditions of a full factorial or the smaller number of treatment combinations for a fractional factorial. With more factors the qualitative variables could, for example, represent the cells of a Graeco-Latin square, which again would be treated as one factor at B levels.

For Case 2, with the same experimental region at each level of the qualitative factor, the product design (13.6) is A- and D-optimum for α and γ in (13.2) with all elements of $v^* = 1/B$, although δ^* will of course depend on the design criterion. Case 3, in which there is some interaction between groups, is not susceptible to general analysis.

In a sense, Kurotschka's Case 1 is not very intersting: designs can be found using the general theory of Chapter 11. Our interest will be in Case 2 which covers many models of practical importance.

***Example* 13.2** Second-order response surface with one qualitative factor (continued)

If the design region for the quantitative factors in (13.3) is the square for which $-1 \leqslant x_i \leqslant 1$ ($i = 1, 2$), the D-optimum continuous design has support at the nine points of the 3^2 factorial for each level of z. The optimum design, which is of product form, has the following design points and weights:

$$4B \text{ corner points } (\pm 1, \pm 1) \qquad 0.1458/B$$

$$4B \text{ centres of sides } (0, \pm 1; \pm 1, 0) \qquad 0.0802/B$$

$$B \text{ centre points } (0, 0) \qquad 0.0960/B. \qquad (13.7)$$

When $B = 1$ this is the D-optimum second-order design for two factors. For general B the design has support at $9B$ design points with unequal weights. The number of parameters is only $B + 5$. Even with a good integer approximation to (13.7), such as repetitions of the 13-trial design formed by replicating the corner points of the 3^2 factorial, the ratio of trials to parameters rapidly becomes intolerable as B increases. In the next section we look for much smaller exact designs. □

13.3 Exact designs

As usual, there is no general construction for exact designs. The design for each value of N has to be calculated individually. In this section we give some examples to demonstrate the features of exact designs and the differences from continuous designs.

Example **13.2** Second-order response surface with one qualitative factor (continued)

To calculate the exact designs for (13.3) a search was made over the nine points of the 3^2 factorial for x_1 and x_2 at each level of the qualitative factor z. There is no constraint on the number of trials n_i at each level except that $n_i \leqslant N$ and $\Sigma_B n_i = N$. Suitable algorithms for the construction of exact designs are described in Chapter 15. Interest was mainly in designs when N is equal to, or just greater than, the number of parameters p. This is not only because of the practical importance of such designs, but also because their structure is furthest from that of the product designs of the continuous theory.

Figure 13.1 shows the D-optimum nine-trial design for model (13.3) when $B = 3$. The number of observations at each level of z is the same, i.e. $n_1 = n_2 = n_3 = 3$, but the design is different for each of the levels. Of course, it does not matter which level of the qualitative factor is associated with which design. One interesting feature is that the projection of the design obtained by ignoring z does not result in the best design when $B = 1$. This, for $N = 9$, is the 3^2 factorial. The best design with such a projection for $B = 3$ has a value of 0.1806×10^{-3} for the determinant $|M(\xi_9)|$, as opposed to 0.1873×10^{-3} for the optimum design—a difference which, whilst real, is negligible for practical purposes. However, for the D-optimum design $d_{ave} = 9.33$ and $d_{max} = 26.84$, whereas for the design which projects into the 3^2 factorial $d_{ave} = 6.45$ and $d_{max} = 16.45$, values which are appreciably better.

A second example, given in Fig. 13.2, exhibits some further properties of the optimum designs. Here $B = 2$ and $N = 13$. The optimum design has five trials at one level and eight at the other, rather than the six to seven division which would be indicated for the continuous product design. Extensive searches of these more equally replicated designs failed to reveal any as good as that of Fig. 13.2. One property of the design is that the designs at each level have a clear structure. Another is that projection of the design yields the 13-trial approximation to the continuous design for $B = 1$ mentioned in §13.2 in which the corner points of the 3^2 factorial are replicated.

The exact design for $N = 18$ and $B = 3$ has the eight-trial design of Fig. 13.2 at one level, with the five-trial design at the others. For the same value of $N = 18$, but with $B = 2$, the design has the eight-trial design of Fig. 13.2 at one level, with two replicates of the five-trial design at the other. The product design of the continuous theory, on the other hand, might lead one to expect two replicates of the 3^2 factorial.

In the examples so far the designs have as their support the points of the 3^2 factorial. If the quantitative factors x can be adjusted more finely than this, one possibility is to search for exact designs over the points of factorials with more

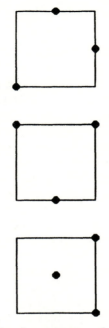

Fig. 13.1. Example 13.2: second-order response surface with one qualitative factor at three levels. *D*-optimum nine-trial design.

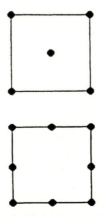

Fig. 13.2. Example 13.2: second-order response with one qualitative factor at two levels. *D*-optimum 13-trial design.

levels. An alternative, which we have found preferable, is to use the adjustment algorithm of §15.7 for the construction of exact designs. An example is shown in Fig. 13.3, where the design of Fig. 13.1 is improved by employment of the adjustment algorithm. However, the increase in D-efficiency is small, from 90.38 to 91.42 per cent. In some environments, particularly where experiments are to be performed by unskilled personnel, this increase may not be worth achieving at the cost of a design in which the factor levels no longer have solely the scaled values -1, 0, and 1. □

Fig. 13.3. Example 13.2: second-order response surface with one qualitative factor at three levels. D-optimum nine-trial design: effect of the adjustment algorithm; ○ design of Fig. 13.1 on points of 3^2 factorial: ● adjusted design points.

Table 13.1 gives the values of the determinants of the information matrices of the designs for Example 13.2 for a variety of values of B and N, as well as for designs with a single quantitative factor. Also given are results from the use of the adjustment algorithm and the D-efficiencies of the designs. Perhaps most informative is the division of the number of trials between the levels of the qualitative factor. These results provide further evidence that for small values of N like these, when product designs are inapplicable, the continuous designs provide little guidance in the construction of exact designs. The algorithms of Chapter 15, such as the BLKL exchange, have to be employed.

13.4 D_β-optimum designs

We now consider the particular case of D_s-optimality when only the parameters β in (13.2) are of interest—a criterion which we call D_β-optimality. Thus only the coefficients of the quantitative factors are of interest. If the information matrix for (13.2) is divided so that $M_{11}(\xi)$ is the information matrix for the qualitative factors, i.e. $M_{11}(\xi_N) = M_{11}(v) = E^T E/N$, and $M_{22}(\xi)$ is likewise the information matrix for the factors x, the criterion requires maximization of the determinant

$$|M_\beta(\xi)| = \frac{|M(\xi)|}{|M_{11}(v)|}$$

$$= |M_{22}(\xi) - M_{12}^T(\xi)M_{11}^{-1}(v)M_{12}(\xi)|. \qquad (13.8)$$

For the product designs (13.5) or (13.6) the value of $M_{11}(v)$ does not depend on the measures δ_i, and so the D- and D_β-optimum continuous designs coincide. As we saw in §13.3, exact D-optimum designs are not of the product form. As a consequence D- and D_β-optimum exact designs are not usually the same. To see how they may differ consider $|M_{11}(v)|$ in (13.8), which is proportional to the product of the sizes of the blocks and so decreases as the block sizes become more unequal, with N remaining fixed. It therefore follows that if the D- and D_β-optimum designs are different, the block sizes for the D_β-optimum design will be less equal. For example, the D-optimum design for $m = 2$, $B = 3$, and $N = 9$ shown in Fig. 13.1 has a $3:3:3$ division between blocks with $|M(\xi_9)| = 0.1873 \times 10^{-3}$ and $|M_\beta(\xi_9)| = 0.6939 \times 10^{-5}$. The D_β-optimum design for the same parameter values is shown in Fig. 13.4. This has the less equal $5:2:2$ division with $|M(\xi_9)| = 0.1427 \times 10^{-3}$ and $|M_\beta(\xi_9)| = 0.7136 \times 10^{-5}$.

Example 13.1 Quadratic regression with a single qualitative factor (continued)

To illustrate a further feature of D_β-optimum exact designs we consider the case with $B = 3$, $N = 7$, and one quantitative factor for which the model is again of second order. The D-optimum design with support $x = -1, 0,$ or 1, shown in Fig. 13.5, has three trials at one level of the qualitative factor and two at the other two levels. For this design $|M(\xi_7)| = 0.7286 \times 10^{-4}$ and $|M_\beta(\xi_7)| = 0.6071 \times 10^{-5}$. The D_β-optimum design, shown in Fig. 13.6, has three trials at two levels and one at the third. Now $|M(\xi_7)|$ has the smaller value of 0.5829×10^{-4}, but the design is better according to D_β-optimality since $|M_\beta(\xi_7)| = 0.6476 \times 10^{-5}$. The single trial in the third block will yield no

Table 13.1. D-optimum exact N-trial designs for a second-order model in m quantitative factors with one qualitative factor at B levels

m	B	n_1	n_2	n_3	p	N	$\|M(\xi_N)\|$	$\|M^A(\xi_N)\|$	D_{eff}	D_{eff}^{AA}
1	2	2	2		4	4	0.156×10^{-1}	0.219×10^{-1}	0.8058	0.8771
1	2	2	3		4	5	0.256×10^{-1}	0.269×10^{-1}	0.9120	0.9234
1	2	3	3		4	6	0.370×10^{-1}	0.370×10^{-1}	1.0000	1.0000
1	3	1	2	2	5	5	0.128×10^{-2}	0.180×10^{-2}	0.7474	0.8001
1	3	2	2	2	5	6	0.309×10^{-2}	0.327×10^{-2}	0.8914	0.9016
1	3	2	2	3	5	7	0.357×10^{-2}	0.394×10^{-2}	0.9175	0.9358
1	3	2	3	3	5	8	0.440×10^{-2}	0.450×10^{-2}	0.9567	0.9610
1	3	3	3	3	5	9	0.549×10^{-2}	0.549×10^{-2}	1.0000	1.0000
2	2	3	4		7	7	0.157×10^{-2}	0.162×10^{-2}	0.9180	0.9222
2	2	4	4		7	8	0.183×10^{-2}	0.188×10^{-2}	0.9384	0.9420
2	2	4	5		7	9	0.193×10^{-2}	0.194×10^{-2}	0.9455	0.9462
2	2	4	6		7	10	0.207×10^{-2}	0.212×10^{-2}	0.9550	0.9583
2	2	5	6		7	11	0.225×10^{-2}	0.229×10^{-2}	0.9665	0.9688
2	2	5	7		7	12	0.236×10^{-2}	0.239×10^{-2}	0.9731	0.9746
2	2	5	8		7	13	0.265×10^{-2}	0.265×10^{-2}	0.9893	0.9893
2	2	6	8		7	14	0.267×10^{-2}	0.267×10^{-2}	0.9903	0.9904
2	2	7	8		7	15	0.255×10^{-2}	0.256×10^{-2}	0.9845	0.9845
2	2	8	8		7	16	0.257×10^{-2}	0.258×10^{-2}	0.9850	0.9855
2	2	8	9		7	17	0.253×10^{-2}	0.253×10^{-2}	0.9828	0.9830
2	2	8	10		7	18	0.253×10^{-2}	0.253×10^{-2}	0.9856	0.9856
2	3	2	3	3	8	8	0.137×10^{-3}	0.145×10^{-3}	0.8693	0.8755
2	3	3	3	3	8	9	0.187×10^{-3}	0.205×10^{-3}	0.9038	0.9142
2	3	3	3	4	8	10	0.246×10^{-3}	0.261×10^{-3}	0.9353	0.9423
2	3	3	3	5	8	11	0.282×10^{-3}	0.295×10^{-3}	0.9514	0.9568
2	3	3	4	5	8	12	0.315×10^{-3}	0.325×10^{-3}	0.9647	0.9685
2	3	5	5	5	8	15	0.363×10^{-3}	0.364×10^{-3}	0.9819	0.9823
2	3	5	5	8	8	18	0.359×10^{-3}	0.359×10^{-3}	0.9806	0.9806

n_j, number of trials at jth level; p, number of parameters; $\|M^A(\xi_N)\|$, determinant resulting from the adjustment algorithm: D_{eff}, D-efficiency.

information about the coefficient of the quantitative factor and so should be omitted. The criterion has provided the best design for the specified values of N and B. However, a better choice would be to take $B=2$ and $N=6$ or 7, depending on the constraints on the experiment. $\square$

An important application of D_β-optimum designs is when the parameters α for the qualitative factors in (13.2) can be regarded as block effects. Then there will be constraints on the number of trials n_i in each block. This special case is the subject of Chapter 14.

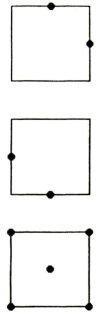

Fig. 13.4. Example 13.2: second-order response surface with one qualitative factor at three levels. D_β-optimum nine-trial design.

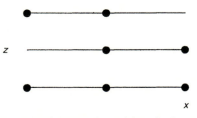

Fig. 13.5. Example 13.1: quadratic regression with a single qualitative factor at three levels. D-optimum seven-trial design.

13.5 Conditions for the optimality of exact designs with both mixture and qualitative factors

Unexpectedly interesting designs with quantitative and qualitative factors arise when the quantitative factors are the q components of a mixture. The extension of Example 12.1 in §12.5 to three wavelengths is one such experiment. As before it is assumed that there is no interaction between the

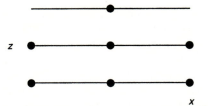

Fig. 13.6. Example 13.1: quadratic regression with a single qualitative factor at three levels. D_β-optimum seven-trial design. The specification $N=7$, $B=3$ has led to a design with only one trial at one of the levels of the qualitative factor. Depending upon circumstances, the design for $B=2$, $N=6$ or $B=2$, $N=7$ would be preferable.

mixture and qualitative factors, so that model (13.2) applies. Further, the qualitative factors will have been expressed so that we need only consider one qualitative factor at B levels.

If the mixture variables can be modelled using second-order canonical polynomials, the model is

$$E(Y) = \sum_{b=1}^{B-1} \alpha_b e_b + \sum_{i=1}^{q} \beta_i x_i + \sum_{i=1}^{q-1} \sum_{j=i+1}^{q-1} \beta_{ij} x_i x_j, \qquad (13.9)$$

where the identifiability condition $\alpha_B = 0$ is imposed. The continuous D-optimum design for (13.9) is again a product design, but now with the lattice design of §12.2 repeated at each level of B. If the number of trials is equal to, or just greater than, the number of parameters in the model, such product designs may again be irrelevant. As in previous sections, we therefore consider reasonably small exact designs. An interesting distinction between these designs for mixtures and those of the previous section is that here it is possible to derive useful theoretical results about the structure and properties of exact designs.

In all cases we shall only consider designs with support at the points of the optimum product design. One important aspect of the design will be the properties of the projection of the design at the various levels of the qualitative factor on to one level. Let this projection design have design matrix F, with $n(s_i)$ trials at support point s_i, where the fitted response is

$$\hat{y}(s_i) = \hat{\beta}^{\mathsf{T}} f(s_i). \qquad (13.10)$$

The projection design is saturated, i.e. there are as many parameters as design points. Therefore it follows from §11.1 that

$$\frac{\mathrm{var}\{\hat{Y}(s_i)\}}{\sigma^2} = \frac{1}{n(s_i)}$$

$$\mathrm{cov}\{\hat{Y}(s_i),\ \hat{Y}(s_j)\} = 0 \qquad (i \neq j). \qquad (13.11)$$

We now show that the determinant of the information matrix of the design is the product of the determinant of the information matrix for the design at the projection level and the determinant of a symmetric matrix R with elements depending both on the number of trials at the same location on the projection level, but at different levels of the qualitative factor, and on the numbers of replicates at the projection level. R is thus purely a counting matrix. This structure greatly simplifies the search for optimum designs.

The D-optimum design maximizes $|M(N)| = |W^T W|$. For the general model (13.2), the determinantal relationship behind (13.8) yields

$$|W^T W| = |F^T F| \times |E^T E - E^T F (F^T F)^{-1} F^T E|, \tag{13.12}$$

a form which emphasizes the importance of the design at the projection level. The determinant of the information matrix of this design is $|F^T F|$, and the elements of $F(F^T F)^{-1} F^T$, which are of the form $f^T(s_i)(F^T F)^{-1} f(s_j)$, are the variances and covariances given by (13.11). Since the covariances are zero, the effect of the indicator variables E is to form sums of these variances. The information matrix $E^T E$ for the indicator variables, is a diagonal matrix $\{n_i\}$, where n_i is the number of trials at the ith level of z.

It is simplest to illustrate the resulting structure of (13.12) for specific values of B. Let O_i denote the set of n_i design points at the ith level of z, let O_{ij} be the subset of points at levels i and j with the same location on the projection level, and let O_{ii} be the set of r_{ii} points at level i with the same location at the projection level as trials from any other level of z. For $B = 3$ with the identifiability condition $\alpha_3 = 0$, $E^T E$ is 2×2 and

$$|W^T W| = |F^T F| \times \left| \begin{Bmatrix} n_1 & 0 \\ 0 & n_2 \end{Bmatrix} - \begin{Bmatrix} \sum_{i \in O_1} n^{-1}(s_i) & \sum_{i \in O_{12}} n^{-1}(s_i) \\ \sum_{i \in O_{12}} n^{-1}(s_i) & \sum_{i \in O_2} n^{-1}(s_i) \end{Bmatrix} \right|. \tag{13.13}$$

For those design points which are not replicated $n(s_i) = 1$, so that (13.13) becomes

$$|M(N)| = |W^T W|$$

$$= |F^T F| \times \begin{vmatrix} r_{11} - \sum_{i \in O_{11}} n^{-1}(s_i) & -\sum_{i \in O_{12}} n^{-1}(s_i) \\ -\sum_{i \in O_{12}} n^{-1}(s_i) & r_{22} - \sum_{i \in O_{22}} n^{-1}(s_i) \end{vmatrix}$$

$$= |F^T F| \times |R|. \tag{13.14}$$

If $B > 3$, the expression for R extends to include the terms obtained by counting shared design points between trials at pairs of levels of z. Otherwise, the structure of $|M(N)|$ is the same. It is the product of two terms, one of which is a counting matrix for the numbers of shared locations at the different levels and the other has only to do with the determinant of the projection matrix.

For $B = 2$ there is appreciable simplification and

$$|M(N)| = |F^T F| \times \left\{ r - \sum_{i=1}^{r} n^{-1}(s_i) \right\} \qquad (13.15)$$

where r $(r \leqslant n_1)$ is the total number of replicated points between the levels. It is clear from (13.15) that to maximize $|M(N)|$ the number of replicated points should be maximized, along with maximization of $|F^T F|$. Then for N less than the number of trials of the product design, the D-optimum N-trial exact design has no more than two replicates at any point on the projection level.

This general result is important for the construction of experiments with both mixture and qualitative factors. The design criterion is seen to be the product of two terms, one dependent only on the replicates of the trials and the other only on the projection design. Thus the generation of an N-trial D-optimum exact design can be carried out in two steps. The first is to generate the exact N-trial D-optimum design for the projection level. The second is to obtain the optimum number of replicated points and their location. The distribution of the non-replicated points does not matter and can be made according to criteria other than D-optimality. There is often appreciable freedom in the positions of the design points and usually several different divisions of the experimental conditions between the levels of the qualitative factor which yield the same value for D-optimality.

The results also have an important interpretation when the qualitative factor is a blocking variable. This is the subject of Chapter 14, but we finish the present chapter with an example of a mixture design in which there are no individual constraints on the numbers of trials n_i at the levels of the qualitative factor.

***Example* 13.3** Second-order model for three-component mixture with qualitative factor at three levels

The 10-trial designs in Fig. 13.7 are all D-optimum for a three-component mixture with a qualitative factor at three levels. Although the division of trials between the levels is different, being $3:3:4$, $2:4:4$, and $2:3:5$, respectively, all give rise to the same design at the projection level. Furthermore, only one pair of points is replicated between levels 1 and 2. It then follows from (13.4) that $|M(10)|$ is the same for each design. To illustrate this we write

$$|M(10)| = |R| \times |F^T F| = |R| \times |M_s(10)|$$

to stress the dependence of the design at the projection level on the number of trials. The determinant for the design at the projection level is $|M_s(10)| = 0.3906 \times 10^{-2}$. According to (13.14)

$$|R| = \begin{vmatrix} 2 - (1/2 + 1/2) & -1/2 \\ -1/2 & 3 - (1/2 + 1/2 + 1/2) \end{vmatrix} = 1.25$$

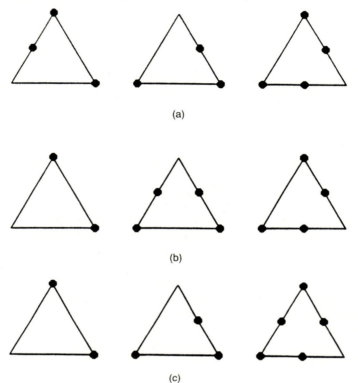

Fig. 13.7. Example 13.1: second-order model for a three-component mixture with a qualitative factor at three levels. D-optimum 10-trial designs. The divisions of the trials between the levels of the qualitative factor are (a) $3:3:4$, (b) $2:4:4$, and (c) $2:3:5$. All three designs give the same value of $|M(\xi_{10})|$.

so that $|M(10)| = 0.4883 \times 10^{-2}$ for all three designs. Because of the identifiability condition $\alpha_3 = 0$, the number of replicates at level 3 does not explicitly appear in this calculation (although it is implicit in the calculations at the projection level). However, relabelling the levels of the qualitative factor can have no effect on the properties of the design. For example, there are two pairs of replicated points between levels 2 and 3 for each of these designs. $\square$

13.6 Further reading

The continuous theory leading to product designs is developed in detail by Kurotschka (1981) and Wierich (1986). Lim *et al.* (1988) consider designs for a general model G in the quantitative factors. Each of the l qualitative factors is constrained to act at two levels. These factors can interact with each other and with submodels G_i of G. The resulting D-optimum continuous designs involve a weighting similar to that of the generalized D-optimum designs of §10.4.

Further discussion and examples of exact designs are given by Donev (1988, Chapters 4 and 5) and Atkinson and Donev (1989).

14
Blocking response surface designs

14.1 Some examples

This chapter, like Chapter 13, is concerned with the design of experiments when the response depends on both qualitative and quantitative factors. For the exact designs of Chapter 13 the number of trials at each level of the qualitative factor is determined by the search algorithm. However, if the qualitative factor corresponds to a blocking variable there may be a specified, and possibly non-constant, number of trials at each level of the qualitative factor. Examples include the number of plots in a field in an agricultural field trial and the number of runs on a chemical plant that can be obtained from a single batch of raw material. This section is concerned with experiments where the number of trials in each block is arbitrary, but specified. Some other situations are discussed in §14.2.

The general model is again

$$E(Y) = W\gamma = E\alpha + F\beta, \tag{14.1}$$

where E is the indicator variable for the B levels of the blocking variable z. The assumption is that there is no interaction between the blocking variable and the quantitative variables. This assumption, customary in the construction of block designs, usually holds in practice. At each level of the blocking variable the number of trials n_i is specified, as is the total number of trials $N = \Sigma n_i$. Because the values of the n_i are specified, a result similar to (13.8) shows that the D-optimum and D_β-optimum designs are the same. If $M_\beta(N)$ is the information matrix for β in (14.1),

$$|M_\beta(N)| = \frac{|M(N)|}{|E^T E|}. \tag{14.2}$$

Since $E^T E$ depends solely on the values of the n_i, which are fixed, the D-optimum design which maximizes $|M(N)|$ in (14.2) will be the same as the D_β-optimum design maximizing $|M_\beta(N)|$. Thus, provided that the block sizes are prespecified, the same design is optimum whether or not the blocking parameters are treated as nuisance parameters.

Example **14.1** Second-order response surface in two blocks

To illustrate the properties of some exact designs for blocking reponse surfaces consider again the second-order response surface model

$$E(Y) = \sum_{i=1}^{B} \alpha_i e_i + \beta_1 x_1 + \beta_2 x_2 + \beta_{11} x_1^2 + \beta_{22} x_2^2 + \beta_{12} x_1 x_2, \qquad (14.3)$$

where support for the design is the 3^2 factorial at each level of the blocking variable.

Figure 14.1 shows, for $N = 10$, the optimum division into two blocks of a range of sizes in which n_1 runs from 2 to 5. Since the labelling of the blocks is not significant, designs for larger values of n_1 are found by interchanging n_1 and n_2. The design for $n_1 = 1$ is not given, since this design, which includes one trial solely to estimate the block parameter for the first block, provides no more information about the response surface than the nine-trial design without blocking. The designs of Fig. 14.2 are for the division of $N = 13$ trials into two blocks.

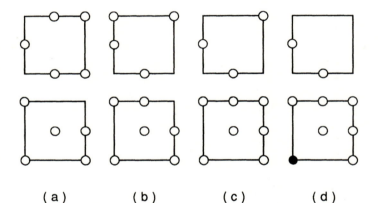

Fig. 14.1. Example 14.1: second-order response surface in two blocks. Designs for $m = 2$, $N = 10$, $B = 2$ ($\bigcirc$ one-trial; $\bullet$ two trials): (a) $n_1 = 5$, $n_2 = 5$; (b) $n_1 = 4$, $n_2 = 6$; (c) $n_1 = 3$, $n_2 = 7$; (d) $n_1 = 2$, $n_2 = 8$.

Some aspects of the structure of these designs deserve comment. For instance, some of the designs can be obtained from one another by sequentially transferring points from one block to another. But generally the designs do not possess this sequential structure. For both $N = 10$ and $N = 13$ the projection of the design on to a single block results in the best N-trial design without blocking. For $N = 10$ this is the 3^2 factorial with any point replicated. The next example serves as a reminder that this property does not hold in general.

The determinants $|M(\xi_N)|$ of the information matrices of the designs shown in Figs 14.1 and 14.2 are given in Tables 14.1 and 14.2. The corresponding values of $|M_\beta(\xi_N)|$, which are relevant when the block effects are treated as

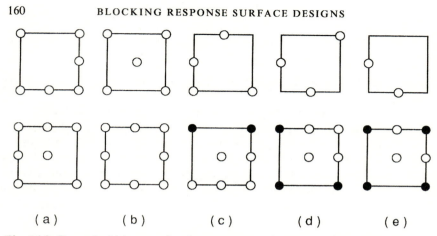

(a)   (b)   (c)   (d)   (e)

Fig. 14.2. Example 14.1: second-order response surface in two blocks. Designs for $m=2$, $N=13$, $B=2$ ($\bigcirc$ one trial; $\bullet$ two trials): (a) $n_1=6$, $n_2=7$; (b) $n_1=5$, $n_2=8$; (c) $n_1=4$, $n_2=9$; (d) $n_1=3$, $n_2=10$; (e) $n_1=2$, $n_2=11$.

nuisance parameters, are also given. Choice of the block sizes for which one or other of the determinants is maximized gives the D- or D_β-optimum designs of Chapter 13. Thus in Table 14.2 the choice $n_1=5$, $n_2=8$ yields the D-optimum design of Fig. 13.2. For other divisions of the trials between the two blocks both $|M(\xi_N)|$ and $|M_\beta(\xi_N)|$ are smaller. However, as was shown by Example 13.1 in §13.4, the D- and D_β-optimum designs do not necessarily coincide when the block sizes are not specified. $\square$

Table 14.1. Example 14.1: second-order response surface in two blocks. Optimum design for $m=2$, $p=7$, $N=10$ (Fig. 14.1)

Block size		$|M(\xi_N)| \times 10^2$	$|M_\beta(\xi_N)| \times 10^4$
n_1	n_2		
1	9	0.0518	0.0576
2	8	0.1142	0.7138
3	7	0.1814	0.8638
4	6	0.2074	0.8642
5	5	0.1978	0.7910

***Example* 14.2** Second-order response surface in three blocks (Example 13.2 continued)

Figures 13.1 and 13.4 show two designs for the division of the nine-trial design

Table 14.2. Example 14.1: second-order response surface in two blocks. Optimum design for $m = 2$, $p = 7$, $N = 13$ (Fig. 14.2)

Block size n_1	n_2	$\lvert M(\xi_N) \rvert \times 10^2$	$\lvert M_\beta(\xi_N) \rvert \times 10^4$
2	11	0.1102	0.5007
3	10	0.1830	0.6099
4	9	0.2223	0.6175
5	8	0.2652	0.6630
6	7	0.2529	0.6023

for the two-factor second-order response surface into three blocks. These designs show less structure than the preceding example of two-block designs. For example, it is not possible to move from the $2:2:5$ design to the $3:3:3$ design by moving just two points. The projection of the $2:2:5$ design is the 3^2 factorial, which is the D-optimum design for $N = 9$ in the absence of blocking. However, we have already seen that this is not the projection of the $3:3:3$ design. □

Example 14.3 Three-component mixture in two blocks

The results of §13.5 for mixture experiments with a qualitative factor indicate that there is great freedom in the distribution of trials between the levels of the factor. Then for a given value of N there will be designs for several different block sizes which all have the same value of $\lvert M(\xi_N) \rvert$. The most striking consequences occur when the design is in two blocks. If the number of points in the second-order lattice is k, the design should have as a projection the N-trial exact D-optimum design for the mixture variables in which $N - k$ of the points are replicated. It was shown in §12.2 that the value of $\lvert M(\xi_N) \rvert$ does not depend on which points are replicated. Then any distribution of the N trials into two blocks for which max $(n_1, n_2) < k$ will yield an exact D-optimum design with the same value of the design criterion. For example, the three nine-trial designs in Fig. 14.3 are equivalent with respect to D-optimality. Those in Figs 14.3(a) and 14.3(b) are chosen to have the same projection and replicated points, but the non-replicated ones are located in a different way. The design in Fig. 14.3(c) has different projection and replicated points, but equal block sizes to that of Fig. 14.3(b). The element of choice in the position of the replicated points in the projection block and of the division of non-replicated points between the blocks makes it possible to achieve improved designs with regard to other criteria. □

The division of a mixture experiment into two blocks illustrates the

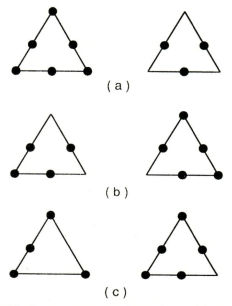

Fig. 14.3. Example 14.3: three-component mixture in two blocks. Some D-optimum designs: (a) $n_1 = 6$, $n_2 = 3$; (b) $n_1 = 4$, $n_2 = 5$, but with the same projection and replication as (a); (c) $n_1 = 4$, $n_2 = 5$ with replicated points and projection different from (b). All designs have the same value of $|M(\xi_N)|$.

simplicity of construction of such designs. Although the designs can readily be constructed by algorithms for exact designs, such as those of the next chapter, a computer is hardly necessary. The same remarks apply to the division of mixture experiments into more blocks. Figure 13.7 gave three different divisions of the 10 trials of a design for a three-component mixture among the three levels of a qualitative factor. Each of these designs is D-optimum for its particular set of block sizes.

The blocked designs of this section, both for second-order response surfaces and for mixture designs, have been found by searching over simple grids in factor space. As Fig. 13.3 shows, a slight improvement in design properties can sometimes be found by the perturbation of moving a small way from the grid. The same is doubtless true of at least some of the designs illustrated in this chapter. For mixture designs the effect of moving away from the lattice will be to destroy the properties which render trivial the construction of designs in blocks. There then seems no alternative but to explore the effect of the perturbation on all optimum designs with support the lattice points. This extra complication may yield a negligible, or no, increase in optimality: for the design of Fig. 13.3 the increase in D-efficiency on moving away from the grid was just over 1 per cent.

14.2 Related problems and literature

In the previous section all block sizes n_i were specified. An important extension is to designs in which there are upper bounds N_i on the number of trials in at least some of the blocks. The numbers of trials per block then satisfy $n_i \leqslant N_i$ with $\Sigma n_i = N$ but $\Sigma N_i > N$, with, perhaps, some block sizes specified exactly, i.e. $n_i = N_i$ for some, but not all, i. Such designs are between those of §14.1 and those of Chapter 13. They can again be constructed by use of an exchange algorithm such as that of §15.7, provided that the constraints on the block sizes can be enforced by the program, as they can in the BLKL algorithm of Appendix A.

Results and methods very similar to those of this chapter are presented by Cook and Nachtsheim (1989), who in addition consider the division into blocks of a specified set of treatments. In one example given by them, an experimenter wished to use a 2^4 factorial design with the inclusion of two centre points, but could only run six trials per day. To guard against the possible effect of variation between days, the 18 trials should be arranged in three blocks of six. With this combination of blocking and design points, it is not possible to estimate all the interactions between the factors. Cook and Nachtsheim take the model to include only the main effects x_i and the two-factor interactions $x_i x_j$. The resulting design is as follows:

block 1	a, d, ab, bc, bd, abcd
block 2	(1), c, ad, abc, acd, bcd
block 3	b, ac, cd, abd, (0), (0).

It is perhaps surprising that both centre points, indicated by (0), occur in the same block.

An advantage of this design is that if blocks turn out not to be important, the full 2^4 factorial is recovered. A general disadvantage is that the stipulated design may not be D-optimum for the particular model, number of trials, and specified blocking structure. Here the centre points, included for the purposes of model-checking, render the design sub-optimum. The best design found by Cook and Nachtsheim for the model with main effects and two-factor interactions in three blocks of six does, however, include all the trials of a 2^4 factorial, with two of the points replicated.

Sometimes an alternative to blocking an experiment is to allow for continuous concomitant variables. The partitioned model is

$$E(Y) = F\beta + Z\gamma \tag{14.4}$$

where the β are the parameters of interest and the z are variables describing the properties of the experimental unit. If the experimental units were people in a clinical trial, the variables in z might include age, blood pressure, and a measure of obesity, variables which would be called prognostic factors. In a field trial, last year's yield on a unit might be a useful concomitant variable.By division of the continuous variables z into categories, blocks could be created which would, in the medical example, contain people of broadly similar age, blood pressure, and obesity. The methods of the previous section could then be applied to find the optimum design for the estimation of β. However, with several concomitant variables, unless the number of units is large, the blocks will tend either to be small or to be so heterogeneous as to be of little use. In such cases there are advantages in designing specifically for the estimation of β in (14.4). Nachtsheim (1989) describes the theory for approximate designs. Exact designs and algorithms are given by Harville (1974, 1975), Jones (1976), Eccleston and Jones (1980), and Jones and Eccleston (1980).

15
Algorithms for the construction of exact D-optimum designs

15.1 Introduction

The difference between exact and continuous designs was stressed in §9.1 and exemplified in Chapters 13 and 14 for models with both qualitative and quantitative factors. For such models it was shown that the continuous product designs are very different from the exact designs with a specified number of trials N. In this chapter we describe algorithms for the construction of exact D-optimum designs. The emphasis is on the methods behind the KL exchange algorithm which was used to calculate many of the examples in this book. A Fortran version of the algorithm and its extension, the BLKL algorithm for blocking response surfaces designs, is given in Appendix A.

The calculation of an optimum design is an optimization problem which can be attacked in several ways. Some methods for continuous designs were described in §§9.4 and 9.5. The next section gives a discussion of problems in the construction of exact designs. Most algorithms contain three phases. In the first phase a starting design of N_0 trials is generated. This is then augmented to N trials and, in the third phase, subjected to iterative improvement. These phases are detailed in §§15.4 and 15.5. Before that, the basic formulae common to the three phases are described in §15.3. The KL exchange, which searches over a grid of candidate points, is presented in §15.6. The adjustment algorithm of §15.7 is one way of moving away from this grid of points. Other algorithms for the construction of exact designs are briefly summarized in §15.8.

15.2 The exact design problem

The exact D-optimum design measure ξ_N^* maximizes

$$|M(N)| = |F^{\mathrm{T}}F| \tag{15.1}$$

where F is an $N \times p$ extended design matrix. For the purposes of this chapter F is a function of m factors which may be quantitative factors continuously variable over a region, qualitative factors, or mixture variables. Because the design is exact, the quantities $N\xi_i$ are restricted to be integer at all design points. The design may include replication so that the number of distinct design points may be less than N.

The optimum design is found by searching over the design region $\mathscr{X}$. For simple problems an analytical solution is sometimes possible, but usually the complexity of the problem is such that numerical methods have to be used. Even for small problems direct search over the whole design region is often not practicable.

Example **15.1** Second-order response surface in two factors

M. J. Box and Draper (1971) use function maximization of the kind described in §9.5 to find exact D-optimum designs for second-order models in $m=2$ and $m=3$ factors with $\mathscr{X}$ a square or cube. When $m=2$ the second-order model has $p=6$ parameters. The exact optimum designs given by them for $N=6, \dots, 9$ are as follows.

$N=6$: $(-1, -1), (1, -1), (-1, 1), (-\alpha, -\alpha), (1, 3\alpha), (3\alpha, 1)$
    where $\alpha = \{4 - \sqrt{13}\}/3 = 0.1315$. Equally optimum designs are obtained by rotation of this design through $\pi/2$, π, or $3\pi/2$.

$N=7$: $(\pm 1, \pm 1), (-0.092, 0.092), (1, -0.067), (0.067, -1)$.

$N=8$: $(\pm 1, \pm 1), (1, 0), (0.082, 1), (0.082, -1), (-0.215, 0)$.

$N=9$: the 3^2 factorial with levels -1, 0, and 1.

The continuous version of this design problem was the subject of §11.5, where it was stated that, for general m, the D-optimum continuous design was supported on subsets of the points of the 3^m factorial; these numerical results show that the general result does not hold for exact designs. The exact designs are illustrated in Fig. 15.1, with some designs rotated to highlight the common structure as N increases. Apart from the design for $N=6$, the designs are very close to fractions of the 3^2 factorial. However, even the design for $N=6$ contains three such points.

The design weights for the continuous D-optimum design for this problem are given by (13.7) with $B=1$. The D-efficiencies of the designs of Fig. 15.1 relative to this design are given in Table 15.1.   □

As the dimension of the problem and the number of factors increase, the time needed to search over the continuous region for the exact design rapidly becomes unacceptable. Following the indication of results such as those of Box and Draper, the search over the continuous region $\mathscr{X}$ is often replaced by a search over a list of candidate points. The list, usually a coarse grid in the experimental region, frequently includes the points of the D-optimum continuous design. In the example above the list might well consist of only the nine points of the 3^2 factorial. The design problem of Example 15.1 is then the combinatorial one of choosing the N out of the available nine points which

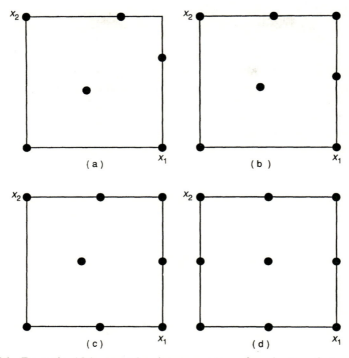

Fig. 15.1. Example 15.1: second-order response surface in two factors. Exact D-optimum designs found by optimization over the continuous region $\mathscr{X}$: (a) $N=6$; (b) $N=7$; (c) $N=8$; (d) $N=9$.

Table 15.1. Example 15.1: second-order response surface in two factors. D-efficiencies of designs of Figs 15.1 and 15.2 found by searching over a continuous square design region and over the 3^2 factorial

N	Points of 3^2 factorial	Continuous square region
6	0.8849	0.8915
7	0.9454	0.9487
8	0.9572	0.9611
9	0.9740	0.9740

maximize $|M(N)|$. In general, the problem is that of selecting N points out of a list of N_e candidate points. Since replication is allowed, the selection is with replacement.

***Example* 15.1** Second-order response surface in two factors (continued)

An alternative to the exact designs given by Box and Draper are the designs of Table 11.6 found by searching over the grid of the 3^2 factorial. The designs are plotted in Fig. 15.2, again with some rotation, now to emphasize the relationship with the designs of Fig. 15.1. Searching over the 25-point grid generated from the 5^2 factorial led to the same designs as searching the 3^m grid.

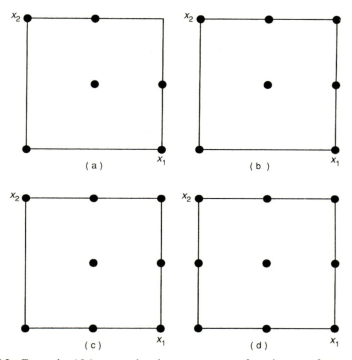

Fig. 15.2. Example 15.1: second-order response surface in two factors. Exact *D*-optimum designs found by searching over the points of the 3^m *and* 5^m factorials: (a) $N=6$; (b) $N=7$; (c) $N=8$; (d) $N=9$.

The *D*-efficiencies of both sets of designs are given in Table 15.1. As can be seen, even for $N=p=6$, little is lost by using the 3^2 grid rather than searching over the continuous design region. Comparison of Figs 15.1 and 15.2 shows that the two sets of designs have the same symmetries and that the

Box–Draper designs involve only slight distortions of the fractions of the 3^m factorial. Use of the finer 5^2 factorial has no advantage over that of the 3^2 grid.

$\square$

Replacement of optimization over $\mathscr{X}$ by a search over the list of candidate points greatly simplifies the search for optimum exact designs. There are also other advantages, the chief of which is that many disparate design problems are reduced to a common optimization structure. Whether the variables are quantitative or qualitative factors or mixture variables, the search reduces to selecting points from a list of length N_c. The incorporation of non-regular or restricted regions, as in Chapter 16, or of several types of variable, as in Chapter 13, introduces no difficulties. Of course, the choice of the list of candidate points is crucial. The list must contain sufficient points to support a good approximation to the exact optimum design over $\mathscr{X}$. But, in cases of doubt, the adjustment algorithm of §15.7, which searches away from the list of candidate points, can be used to check that a good design has been found.

15.3 Basic formulae for exchange algorithms

Numerical algorithms for the construction of exact D-optimum designs by searching over a list of N_c candidate points customarily involve the iterative improvement of an initial N-trial design. In general, the algorithms are attempting to find the maximum of a surface with many local extrema. Many of the algorithms cannot be guaranteed to find anything more than a local optimum. The probability of finding the global optimum can be increased by repeating the search several times from different starting designs, the generation of which often includes a random component. The probability of finding the global optimum can also be increased by the use of a more thorough search algorithm for improvement of the initial design. With a long list of candidate points and a model with several parameters, complicated search algorithms can readily consume appreciable computer time and space. The problem is discussed by Atkinson and Donev (1989). In general, for restricted resources, there needs to be a balance between the number of starting designs and the complexity of the search algorithm. The procedure described in §15.6 uses a simple search, but requires several starts or 'tries' (Mitchell 1974a) to have a reasonable chance of locating the best design, or one close to it.

The initial design of size N can be constructed sequentially from a starting design of size N_0, either by the addition of points if $N_0 < N$, or by the deletion of points if $N_0 > N$. Improvement of the design in the third phase is made by an exchange in which points in the design are replaced by those selected from the candidate list, with the number of points N remaining fixed. The common structure is that the algorithms add a point x_l to the design, delete a point x_k

from it, or replace a point x_k from the design with a point x_l from the list of candidate points. The choice of the points x_k and x_l depends on the variance of the predicted response at these points, the determinant of the information matrix, and the values of elements of its inverse. The search strategy is determined by the algorithm, several of which are described in the next sections. In this section we give the formulae which provide updated information at each iteration.

Let n $(n \geqslant 0)$ be the number of iterations already performed. To combine the sequential and interchange steps in a single formula we require constants c_k and c_l such that

(1) $c_l = (N+1)^{-1}$, $c_k = 0$ if a point x_l is added to the design;

(2) $c_k = (N+1)^{-1}$, $c_l = 0$ if the design point x_k is deleted;

(3) $c_k = c_l = (N+1)^{-1}$ if the design point x_l is exchanged with the point x_k.

Let the vector of model terms which form the rows of F at these points be $f_k^T = f^T(x_k)$ and $f_l^T = f^T(x_l)$. The following formulae then give the relation between the information matrices, their determinants, and the inverses of the information matrices at the nth and $(n+1)$th iterations:

$$M(\xi_{n+1}) = \frac{1-c_l}{1-c_k} M(\xi_n) + \frac{1}{1-c_k} (c_l f_l f_l^T - c_k f_k f_k^T) \qquad (15.2)$$

$$|M(\xi_{n+1})| = \left[\left\{ 1 + \frac{c_l}{1+c_l} d(x_l, \xi_n) \right\} \left\{ 1 - \frac{c_k}{1-c_k} d(x_k, \xi_n) \right\} \right.$$

$$\left. + \frac{c_k c_l}{(1-c_l)^2} d^2(x_k, x_l, \xi_n) \right] \left(\frac{1-c_l}{1-c_k} \right)^p |M(\xi_n)| \qquad (15.3)$$

$$M^{-1}(\xi_{n+1}) = \frac{1-c_l}{1-c_k} \left\{ M^{-1}(\xi_n) - \frac{M^{-1}(\xi_n)AM^{-1}(\xi_n)}{qz + c_k c_l d^2(x_l, x_k, \xi_n)} \right\} \qquad (15.4)$$

where

$$d(x_l, x_k, \xi_n) = f_l^T M^{-1}(\xi_n) f_k$$

$$q = 1 - c_l + c_l d(x_l, \xi_n)$$

$$z = 1 - c_l - c_k d(x_k, \xi_n)$$

and

$$A = c_l z f_l f_l^T + c_k c_l d^2(x_l, x_k, \xi_n)(f_l f_k^T + f_k f_l^T) - c_k q f_k f_k^T. \qquad (15.5)$$

The computer implementation of these formulae requires care: updating the design and the inverse of its information matrix, in addition to recalculation of the variances at the design points, can consume computer time and space. However, some advice can be given.

1. The inverse of the information matrix is symmetric, so that only a triangular matrix need be stored.

2. Either the list of candidate points or the whole extended design matrix can be kept. In the former case the $f_j(j = 1, \ldots, N_c)$ must be calculated each time they are needed. Compared with storing the extended design matrix this takes more time, but saves space. The choice depends on both the computer and the size of problem.

15.4 Sequential algorithms

An exact design for N trials can be found either by the sequential addition of trials to a smaller design, or by the sequential deletion of trials from a larger design. If necessary, the exact N-trial design can then be improved by the methods described in the next two sections. The formulae of §15.3 apply for either addition or deletion.

1. *Forwards procedure* Given a starting design with N_0 trials, the N-trial exact design $(N > N_0)$ is found by application of the algorithm illustrated in §11.2, i.e. each trial is added sequentially at the point where the variance of the predicted response is greatest:

$$d(x_l, \xi_i) = \max_x d(x, \xi_i) \qquad (N_0 \leqslant i \leqslant N). \tag{15.6}$$

As was shown in §9.4, continuation of this process leads, as $N \to \infty$, to the D-optimum continuous design ξ^*. The exact design yielded by the application of (15.6) can then be regarded as an approximation to ξ^* which improves as N increases.

2. *Backwards procedure* This starts from a design with $N_0 \gg p$ and proceeds by the sequential deletion of points with low variance. At each iteration we delete the design point x_k at which the variance of the predicted response is a minimum:

$$d(x_k, \xi_i) = \min_x d(x, \xi_i) \qquad (N \leqslant i \leqslant N_0). \tag{15.7}$$

As (15.3) shows, the decrease in the value of the determinant of the information matrix at each iteration is the minimum possible. Often the list of candidate points is taken as the starting design for this procedure. However, if N is so large that the N-trial design might contain some replication, the starting design could contain two or more trials at each candidate point.

A common feature of both these one-step-ahead procedures is that they do not usually lead to the best exact N-trial design. The backwards procedure is clumsy in comparison with the forwards procedure because of the size of the starting design. The performance of the forwards procedure can be improved by using different starting designs. In order to generate a series of starting designs Galil and Kiefer (1980) suggest selecting j points $(0 < j < p)$ at random from the candidate set. For this design $F^T F$ will be singular. However, the $N \times N$ matrix FF^T can be used instead with the forwards procedure until the design is no longer singular. Thereafter (15.6) is used directly. Thus different runs of the algorithm will produce a variety of exact N-trial designs, the best of which will be selected. The KL exchange algorithm described in §15.6 uses another method to obtain a variety of starting designs. The method, which relies upon the regularization of singular $F^T F$, can be adapted to allow division of the trials of the design into blocks of specified size and so provide the designs of Chapter 14.

In some practical situations, particularly when the number of trials is appreciably greater than the number of parameters, the forwards sequential procedure yields a satisfactory design. In others, the design will need to be improved by one of the non-sequential methods of the next two sections.

15.5 Non-sequential algorithms

The non-sequential algorithms are intended for the improvement of an N-trial exact design. This is achieved by deleting, adding, or exchanging points, according to the rules of the particular algorithm, to obtain an improved N-trial design. Because the procedures are non-sequential, it is possible that the best design of N trials might be quite different from that obtained for $N - 1$ or $N + 1$ trials.

Van Schalkwyk (1971) proposed an algorithm which at each iteration deleted the point x_k from the design to cause minimum decrease in the determinant of the information matrix as in (15.7). The N-trial design is then recovered by adding the x_l which gives a maximum increase of the determinant, thus satisfying (15.6) for $i = N - 1$. Progress ceases when the same point is deleted and then re-entered. Mitchell and Miller (1970) and Wynn (1970) suggest a similar algorithm in which the same actions are performed, but in the opposite order: first the point x_l is added and then the point x_k is deleted, with the points for addition and deletion being decided by the same rule as in van Schalkwyk's algorithm.

The idea of alternate addition and deletion of points is extended in the DETMAX algorithm (Mitchell 1974a) to 'excursions' of various sizes in which a chosen number of points is sequentially added and then deleted. The size of the excursions increases as the search proceeds, usually up to a maximum of six. Of course, a size of unity corresponds to the algorithm described at the end of the previous paragraph. Galil and Kiefer (1980) describe computational improvements to the algorithm and generate D-optimum exact designs for second-order models in three, four, and five factors.

DETMAX, like the other algorithms described so far, separates the searches for addition and deletion. The two operations are considered together in the exchange algorithm suggested by Fedorov (1972, p. 164): at each iteration the algorithm evaluates all possible exchanges of pairs of points x_k from the design and x_l from the set of candidate points. The exchange giving the greatest increase in the determinant of the information matrix, assessed by application of (15.3), is undertaken: the process continues as long as an interchange increases the determinant. As one way of speeding up the algorithm, Cook and Nachtsheim (1980) consider each design point in turn, perhaps in random order, carrying out any beneficial exchange as soon as it is discovered. They call the resulting procedure a modified Fedorov exchange. Johnson and Nachtsheim (1983) further reduce the number of points to be considered for exchange by searching over only the k $(k < N)$ design points with lowest variance. The algorithm of the next section generalizes Johnson and Nachtsheim's modification of Fedorov's original exchange procedure.

15.6 The KL and BLKL exchange algorithms

As the exchange algorithms for exact designs are finding local optima of functions with many extrema, improvement can come from an increased number of starting designs as well as from more precise identification of local optima. Experience suggests that, for fixed computational cost, there are benefits from a proper balance between the number of tries and the elaborateness of each search. For example, Fedorov's exchange algorithm is made slow by the large number of points to be considered at each iteration (a maximum of $N(N_c - 1)$ in the absence of replication in the design) and by the need to follow each successful exchange by updating the design, the covariance matrix $M^{-1}(\xi)$, and the variance of the predicted values at the design and candidate points. The thoroughness of the search contributes to the success of the algorithm. However, the search can be made faster by noting that the points most likely to be exchanged are design points with relatively low variance of the predicted response and candidate points for which the variance is relatively high. This idea underlies the KL exchange and its extension to blocking, which we have called BLKL.

The algorithm passes through three phases:

(1) generation of the starting design;
(2) sequential generation of the initial N-trial design;
(3) improvement of the N-trial design by exchange.

Sometimes there may be points which the experimenter wishes to include in the design. The purpose might be to check the model, or they might represent data already available. The first phase starts with q_1 ($q_1 \geqslant 0$) such points. The random starts to the search for the optimum comes from choosing q_2 points at random from the candidate set, where q_2 is itself a randomly chosen integer $0 \leqslant q_2 \leqslant \min(N - q_1, [p/2])$ and $[a]$ is the integer part of a.

The initial N-trial design is completed by sequential addition of those $N - (q_1 + q_2)$ points which give maximum increase to the determinant of the information matrix. For $N < p$ the design will be singular and is regularized, as in (10.7), by replacement of $F^T F$ by $F^T F + \varepsilon I_p$, where ε is a small number, typically between 10^{-4} and 10^{-6}. If the design is to be laid out in blocks, the search for the next design point is confined to those parts of the candidate set corresponding to non-full blocks.

In the third phase the exchange of points x_k from the design and x_l from the candidate list is considered. As in other algorithms of the exchange type, that exchange is performed which leads to the greatest increase in the determinant of the information matrix. The algorithm terminates when there is no longer any exchange which would increase the determinant. The points x_k and x_l considered for exchange are determined by parameters K and L such that

$$1 \leqslant k \leqslant K \leqslant N$$

and

$$1 \leqslant l \leqslant L \leqslant N_c - 1.$$

The point x_k is that with the kth lowest variance of prediction among the N design points, and x_l has the lth highest variance among the N_c candidate points. If blocking is required, the orderings of points should theoretically be over each block with exchanges limited to pairs within the same block. However, we have not found this refinement to be necessary. Whether blocking is required or not, the q_1 points added at the beginning of the starting procedure are not considered for exchange.

When $K = N$ and $L = N_c - 1$, the KL exchange coincides with Fedorov's procedure. However, by choosing $K < N$ and $L < N_c - 1$, the number of pairs of points to be considered at each iteration is decreased. Although this must diminish the probability of finding the best possible exact design at each try, the decrease can be made negligible if K and L are properly chosen. The advantage is the decrease in computational time.

There are two possible modifications to this algorithm:

(1) to make all beneficial exchanges as soon as they are discovered, updating the design after each exchange;

(2) to choose the K design points and L candidate points at random rather than according to their variances.

The first modification brings the algorithm close to the modified Fedorov procedure of Cook and Nachtsheim (1980) when $K=N$ and $L=1$ and becomes the K exchange (Johnson and Nachtsheim 1983) for $0<K<N$ and $L=1$. Our extension to include the choice of L provides extra flexibility.

The second modification, that of the random choice of points, could be used when further increase in the speed of the algorithm is required. In this case the variance of the predicted value is calculated only for the $K+L$ points, rather than for all points in the design and the candidate set. In the unmodified algorithm this larger calculation is followed by ordering of the points to identify the $K+L$ for exchange. This modification should yield a relatively efficient algorithm when the number of design or candidate points is large.

The best values of K and L depend, amongst other variables, on the number of factors, the degrees of freedom for error $v=N-p$, and the number of candidate points N_c. For example, when $v=0$, the variance of the predicted response at all design points is the same: there is then no justification for taking $K<N$. However, as v increases, the ratio K/N should decrease. The best value of L increases with the size of the problem, but never exceeded $N_c/2$ in the examples considered by Donev (1988). In most cases values of K and L much smaller than these limits are sufficient, particularly if the number of tries is high. In an example reported in detail by Donev, the unmodified Fedorov algorithm was used for repeated tries on two test problems in order to find the values of K and L required so that the optimum exchange was always made. In none of the tries would taking K or L equal to unity have yielded the optimum exchange—usually much larger values were necessary.

Since the values of K and L are problem dependent, they are included as parameters in the algorithm of Appendix A. A little experimentation with these two factors should rapidly yield a satisfactory set of values.

15.7   An adjustment algorithm

The KL exchange searches over a list of candidate points. For quantitative factors the list often consists of a coarse grid, such as the points of a 3^m factorial. But, as Example 15.1 showed, even if these comprise the points of support of the optimum continuous design, they may not support the optimum exact design. In this section a description is given of the adjustment algorithm, which searches away from the candidate list in the neighbourhood of a good exact design. An example of the application of the algorithm has already been given in Fig. 13.3.

As a result of applying DETMAX or the KL exchange algorithm a design is obtained at a local optimum for the given candidate set. The adjustment algorithm is a method of finding this local optimum more precisely by adjusting the design points. At each of the N points of the proposed exact design the effect of moving the design point a small amount along each factor axis is calculated. The effect of the change on the determinant $|M(\xi_N)|$ is calculated from (15.3). If an increase is possible, the single point for which it is greatest is changed and the process is repeated until progress ceases. At the most there are $2mN$ perturbations per stage, less if any of the design points are on the boundary of $\mathscr{X}$ or if there is any replication in the design.

More formally, let s_k be a step length $(k = 1, \ldots, m)$ and let SM be the minimum step length to be considered. The steps of the algorithm are as follows.

1. Calculate from (15.3) the effects r_{ij} $(i = 1, \ldots, N, j = 1, \ldots, 2m)$ of the perturbations about each of the N design points, where r_{ij} is the ratio of determinants. For each perturbed point the kth co-ordinate differs from that of the design point according to the rule

$$x'_k = x_k \pm s_k \qquad (k = 1, \ldots, m).$$

   Points outside the design region are omitted.

2. Find $r^* = \max r_{ij}$. Let the indices be i^* and j^*. If the determinant has increased, the ratio r^* will be greater than unity.

3. If $r^* > 1$, exchange the design point i^* for the value of x'_k at (i^*, j^*) and return to Step 1. Otherwise go to Step 4.

4. $s_k = s_k/2$ $(k = 1, \ldots, m)$.

5. If $s_k \geqslant$ SM $(k = 1, \ldots, m$ for at least one $k)$, go to Step 1. Otherwise stop.

The details of the step-length reduction in Step 4 do not seem to be critical. In applications to response surface designs with all factors scaled between -1 and 1, an alternative employed was to take a first value of 0.1 for s_k, followed by a second finer search with $s_k = 0.01$. This finer search never achieved a significant improvement to the design found with $s_k = 0.1$.

Figure 13.3 shows the way in which application of the adjustment algorithm moves the design away from the grid of candidate points. With the exception of the design for $N = 6$, application of the algorithm to the exact designs for the second-order response surface of Example 15.1 produces the designs of Fig. 15.1 from those of Fig. 15.2. The results of Table 15.2 show the effect of the adjustment algorithm on the properties of the designs of Table 11.6 for second-order models, including those of Example 15.1, in up to five factors. The algorithm increases the D-efficiency in all examples except one. The effect of the algorithm on the D-optimality of designs for blocking second-order response surfaces is shown in Table 13.1. As is indicated by these tables and by comparison of Fig. 15.1 with Fig. 15.2, the effect of the adjustment is greatest when N is equal to, or just greater than, p.

Although the algorithm is intended to improve D-optimum designs, it is interesting to compare the designs on the basis of G- and V-optimality: it

Table 15.2. Second-order polynomial in m factors: cubic experimental region. Effect of the adjustment algorithm on the exact D-optimum designs of Table 11.6 (Table 11.8 extended)

m	N	p	D_{eff}	D_{eff}^A	G_{eff}	G_{eff}^A	d_{ave}	d_{ave}^A
2	6	6*	0.8040	0.8915	0.1765	0.5305	7.70	6.04
2	7	6	0.9454	0.9487	0.6122	0.6310	5.74	5.67
2	8	6	0.9572	0.9611	0.6000	0.6661	5.38	5.28
3	10	10	0.8631	0.8925	0.2904	0.5507	11.22	8.88
3	14	10	0.9759	0.9759	0.8929	0.8929	7.82	7.82
3	15	10	0.9684	0.9701	0.7752	0.7819	8.85	8.82
3	16	10	0.9660	0.9671	0.7418	0.7375	8.81	8.78
3	18	10	0.9717	0.9722	0.6817	0.7117	9.22	9.20
3	20	10	0.9779	0.9784	0.8258	0.8347	8.93	8.92
4	15	15	0.8700	0.8756	0.4522	0.5386	19.28	16.87
4	18	15	0.9311	0.9345	0.5114	0.5203	14.50	14.46
4	24	15	0.9670	0.9688	0.6640	0.7232	14.16	14.18
4	25	15	0.9773	0.9778	0.6890	0.7046	14.07	14.07
4	27	15	0.9815	0.9819	0.6983	0.6980	13.89	13.88
5	21	21	0.9055	0.9085	0.4471	0.4563	26.62	26.39
5	26	21	0.9519	0.9525	0.6705	0.6731	19.86	19.86
5	27	21	0.9539	0.9555	0.6494	0.6498	19.71	19.69

D_{eff}^A, G_{eff}^A, and d_{ave}^A are properties of adjusted designs.
*The initial design yielding this result was not that of Table 11.6. The design points were 2 3 4 5 7 9.

might be expected that the more even filling of the design region by moving away from a coarse grid would have an appreciable effect on the variance of the predicted response. Because of the computational cost of finding the maximum and average variances over the whole of $\mathscr{X}$, the variance was, as in Chapter 11, calculated only over the points of a grid. For the second-order response surface designs of Table 15.2 the search was over the points of the 5^m factorial. For these applications of the adjustment algorithm the initial step length in all directions was $s_k = 0.08$, with the lower bound SM $= 0.02$. The results show that the procedure increases the G-efficiency of the designs, most noticeably when $N = p$. For comparison of the average variances small values are desirable. Table 15.2 shows that, in most cases, the adjustment algorithm leads to an appreciable improvement in V-optimality, whereas in a few cases there is a slight increase in average variance.

The adjustment algorithm is very simple. Of course, more complicated search strategies are possible. But limited experience suggests that these add computational complexity without yielding improved designs. Although the search is not particularly efficient, it is only employed on the single best design found by the exchange algorithm. After 100 tries of a complicated search, for example, the computational efficiency of the final single search using the adjustment algorithm is not critical. However, it is worthwhile, for example, for blocked response surfaces designs with few trials per block, where the number of trials and parameters are comparable.

15.8 Other algorithms and further reading

In principle, it is easy to modify the exchange, and other non-sequential, algorithms for criteria other than *D*-optimality. For example, Welch (1984) describes how DETMAX can be adapted to generate *G*- and *V*-optimum designs.

Neither DETMAX nor the exchange algorithms are guaranteed to find the globally optimum design. However, globally optimum designs can be found, at the cost of increased computation, by the use of standard methods for combinatorial optimization problems. Bohachevsky *et al.* (1986) use simulated annealing to construct an exact design for a non-linear problem in which replication of points is not allowed. The search is not restricted to a list of candidate points. In the application of simulated annealing reported by Haines (1987), *D*-, *G*-, and *V*-optimum exact designs were found for a variety of polynomial models. Comparisons with the *K* exchange of Johnson and Nachtsheim (1983) indicate that the exchange algorithm is preferable for *D*-optimality and the annealing algorithm is preferable for *G*-optimality, but that there is little to choose between the algorithms for *V*-optimality.

A second general method of search is that of branch and bound, which is applied by Welch (1982) to finding globally *D*-optimum designs over a list of candidate points. The method guarantees to find the optimum design. However, the computational requirements increase rapidly with the size of the problem.

16
Restricted design regions

16.1 Introduction

A major advantage of a general theory such as that of optimum experimental design is the provision of methods for the solution of non-standard problems. In this chapter we illustrate this consequence of the theory through the construction of designs for non-standard design regions. For response surface models these are regions which are neither cubical nor spherical and so are not covered by the results of Chapter 11. For mixture designs, the non-standard regions depart from the simplex for which designs were derived in Chapter 12.

A description of a variety of non-standard design problems is given by Snee (1985). His major theme is the usefulness of the computer in the design and conduct of a wide variety of experiments. One area of application is the construction of D-optimum designs for irregular experimental regions. He gives seven examples of such regions and discusses desirable properties of designs, but does not calculate the D-optimum designs. The examples giving rise to non-standard regions are listed briefly below before we consider four examples in more detail.

1. *Polymer density* A central composite design over a circular design region was used for a study in which the factors were annealing time and annealing temperature. However, the highest temperature was so great that the polymer melted. The circular design region was therefore made non-standard by the introduction of an upper bound on one of the factors.

2. *Petroleum fractionation* Again there are two process variables: the amount of toluene and the solvent-to-solute ratio. The process is known to be uneconomic at low levels of both variables and technically infeasible at high levels of both. The resulting design region is a parallelogram. Example 16.2 below, taken from motor manufacture, is a similarly constrained problem.

3. *Product stability* If the two variables of interest are product age and the amount of time the product has spent in the customer's warehouse, the design region is triangular.

4. *Fertilizer trials* High levels of two or more fertilizers can produce conditions under which plants die, perhaps because the concentration of ions in the soil becomes so high that osmosis fails to provide the plant with water. Imposition of the constraint $x_1 + x_2 < c$ avoids simultaneous high

values of both factors. The design region is a square with one corner removed, as in Fig. 2.2(d).

5. *Ortho-xylene oxidation* In this extension of the ideas in examples 2 and 4, the three-factor region is constrained by two non-parallel hyperplanes (Juusola *et al.* 1972).

6. *Synthetic fibre manufacture* In an experiment on the physical properties of a synthetic fibre it was found that, because of physical constraints on the system, it was not possible to run all the points of a full $3^4 \times 4$ factorial in the process variables. Kennard and Stone (1969) selected a subset of the points of the full factorial.

7. *Aerosol propellant studies* In this series of mixture experiments there were several linear constraints on the design region, only some of which led to regions which were again simplexes and so could be handled using the pseudo-components of §12.3.

Example 16.3 is one such case of a mixture experiment with a non-standard region.

There are two general ways to proceed. One is to take a 'standard' design and to try to adapt it to the non-standard region. The second is to use the algorithms for continuous or exact optimum designs to find the best design for the specific region. Clearly, we favour the second possibility which involves little more work than finding the optimum design for a standard region. For exact designs, the only complication may arise in finding the set of candidate points. In the example 6 the candidate set was taken as the points of a five-factor factorial from which infeasible combinations were eliminated. This is an easy way of finding a set of feasible design points. However, particularly if the constraints can be expressed analytically, it will often be desirable to add candidate points which lie on the constraints. Otherwise too sparse a candidate set may result. This point is illustrated by Example 16.2.

In the remainder of this chapter we illustrate the construction of *D*-optimum designs for irregular design regions. Two examples are of response surface designs and two are of designs for mixtures.

16.2 Response surfaces

As a first example of a design for a non-standard region we take a simple problem due to Wynn (1970) which he used to illustrate the sequential construction of *D*-optimum designs.

Example 16.1 First-order model
For the first-order model in two factors

$$E(Y) = \beta_0 + \beta_1 x_1 + \beta_2 x_2$$

the customary design region is a circle or a square. Suppose instead that $\mathcal{X}$ is as shown in Fig. 16.1. The vertices B $(-1, 1)$, C $(1, -1)$, and D $(-1, -1)$ lie on three corners of the unit square, but the other point A has been moved out to $(2, 2)$.

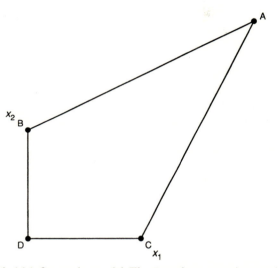

Fig. 16.1. Example 16.1: first-order model. The D-optimum continuous design puts the unequal design weights of Table 16.1 at the four vertices of $\mathcal{X}$.

Table 16.1. Example 16.1: first-order model. D-optimum continuous design and exact design for $N = 32$ for the non-regular design region of Fig. 16.1

Design point	x_1	x_2	Design weight
A	2	2	10/32
B	−1	1	9/32
C	1	−1	9/32
D	−1	−1	4/32

Finding the D-optimum design for this region is straightforward since, with the first-order model, the points of support of the optimum design must be at the vertices of $\mathcal{X}$. Searching over these vertices gives the optimum design of Table 16.1. In this design the extreme point A receives a weight of 10/32, whereas the point diagonally opposite has a weight of only 4/32. The remaining two points have weights of 9/32 each.   □

To find the design Wynn used the iterative algorithm for the construction of D-optimum designs in which one trial is added at a time at the point of maximum variance. The optimum continuous design of Table 16.1 was obtained for $N = 32$. Usually the construction of the design is not so straightforward: an algorithm such as BLKL has to be used if an exact design is required.

***Example* 16.2** Performance of an internal combustion engine

In this example, which is a simplified version of a problem which arises in testing car engines, there are two factors, spark advance x_1 and torque x_2. Although both are independently variable, so that a square design region is theoretically possible, certain factor combinations have to be excluded, either because the engine would not run under these conditions, or because it would be damaged or destroyed. The excluded combinations covered trials when the two factors were both at either high or low levels. The resulting pentagonal design region is shown in Fig. 16.2.

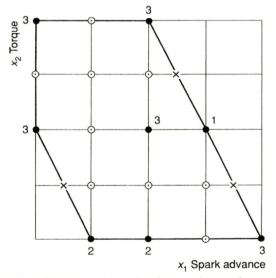

Fig. 16.2. Example 16.2: internal combustion engine performance: ● 20-trial D-optimum design with number of trials at each point; ○ unused candidate points; × second-stage candidate points.

It is known from previous experiments that a second-order model in both factors is needed to model the response and that 20 trials give sufficient accuracy. The exact D-optimum design was found using the BLKL algorithm where the initial candidate set comprised the 17 points of the 5^2 factorial which satisfied the constraints.

The candidate set is shown by circles in Fig. 16.2. Solid circles denote the points of the optimum design, and open circles denote those members of the candidate set which were not used. The resulting D-optimum design has support on eight out of the total 17 points. Because of the oblique nature of the constraints, some of the candidate points are distant from the constraints. The three extra candidate points marked by crosses were therefore added to give a denser candidate set. Repeating the search over this enlarged set of 20 points yielded the same optimum design as before with eight support points. The extra candidate points were not used.

There are two general points about this example. The first is that exclusion of candidate points on a grid which lie outside the constraints of the design region may result in too sparse a candidate set. (In our example this was not the case.) More important is the design of Fig. 16.2. This has eight support points. Perhaps something almost as efficient could have been found by an inspired distortion of the nine-trial D-optimum design for the square region. Our experience suggests that this is the initial reaction of experimenters faced with constrained design problems yielding irregular regions. But even if such inspiration did descend on the experimenter, it would not be needed. Once a suitable candidate set has been defined, such algorithms as BLKL work independently of any symmetries or structure in the region: an irregular region is treated no differently from one for a standard problem.   □

16.3 Mixture experiments

In this section designs are found for two constrained mixture problems which cannot be solved using the pseudo-components of §12.3.

Example 16.3 Linear constraints

The three components x_i of a mixture are subject to the constraints

$$0.2 \leqslant x_1 \leqslant 0.7$$
$$0.1 \leqslant x_2 \leqslant 0.6$$
$$0.1 \leqslant x_3 \leqslant 0.6,$$

as well as the usual mixture constraints. The resulting hexagonal design region is shown in Fig. 16.3. The figure also shows the points of the extreme vertices design constructed following the method of McLean and Anderson (1966) which consists of all vertices, centres of edges, and the centre point, 13 points in all.

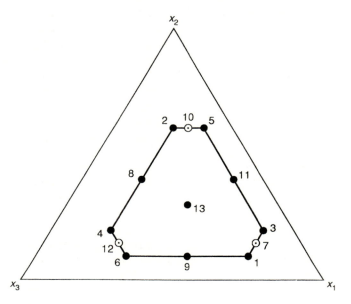

Fig. 16.3. Example 16.3: mixture experiment with linear constraints. D-optimum design with points numbered as in Table 16.2: ● design points; ○ unused candidate points.

Suppose that the model is the second-order canonical polynomial

$$E(Y) = \beta_1 x_1 + \beta_2 x_2 + \beta_3 x_3 + \beta_{12} x_1 x_2 + \beta_{13} x_1 x_2 + \beta_{23} x_2 x_3,$$

with six parameters. The extreme vertices design could be used. However, this design takes no account of the model, merely reflecting the structure of $\mathcal{X}$. Further, the design can be expected to be inefficient for the second-order model as the number of design points would be appreciably larger than the number of parameters. We therefore take the points of the extreme vertices design as a list of candidate points for the construction of exact designs.

The results of using the BLKL algorithm to find moderate-sized exact designs are given in Table 16.2. For $N = 10$ the design puts one trial at each point of the candidate set except for the centres of the short edges. The design is shown in Fig. 16.3, where the open symbols represent the unused candidate points. For $N = 20$ and $N = 30$ this pattern is repeated two and three times respectively. For $N = 35$ this repetition is not possible: the design in Table 16.2 puts only two trials at the centre point, as a result of which there is a symmetrical structure in the other points. Points at the centres of the long sides and at one end of the short sides each receive four trials, with three trials at the other end of each short side. Designs with equal efficiency can be found by interchanging the points on short sides receiving three and four trials.

Table 16.2. Example 16.3: mixture experiment with linear constraints. Exact D-optimum designs for the non-standard design region in Fig. 16.3 (second-order model)

Point number	x_1	x_2	x_3	Exact designs (BLKL) $N=10$	$N=20$	$N=30$	$N=35$	Sequential design $N=997$
1	0.7	0.1	0.2	1	2	3	4	100
2	0.2	0.6	0.2	1	2	3	4	100
3	0.7	0.2	0.1	1	2	3	3	100
4	0.2	0.2	0.6	1	2	3	3	100
5	0.3	0.6	0.1	1	2	3	3	100
6	0.3	0.1	0.6	1	2	3	4	100
7	0.7	0.15	0.15					
8	0.2	0.4	0.4	1	2	3	4	109
9	0.5	0.1	0.4	1	2	3	4	109
10	0.25	0.6	0.15					
11	0.5	0.4	0.1	1	2	3	4	109
12	0.25	0.15	0.6					
13	0.4	0.3	0.3	1	2	3	2	70

The design for $N=35$ does not give quite as efficient a design measure as the equi-replicated design, although the difference is less than 0.1 per cent. However, the design does serve as a further reminder that exact designs do depend, sometimes in an unpredictable way, on the specific value of N. To obtain some idea about the continuous D-optimum design we used the first-order algorithm adding design points where the variance $d(x, \xi_N)$ is a maximum. The starting design was the equi-replicated 10-point design of Table 16.2. As we saw in §11.2, this algorithm does not converge monotonically towards the optimum but passes through good balanced designs, which are then distorted by the sequential addition of design points. In this example a particularly good design was found for 997 trials, with $\bar{d}(\xi) = 6.0068$, as opposed to six for the continuous D-optimum design. The structure is that the centre point has the lowest weight, the centres of the long sides have the highest weight, and the remaining weight is equally distributed over the end-points of the short sides of the design region.

This example thus illustrates two features of mixture designs for constrained regions. One is that not all the candidate points of the extreme vertices design are required. The second is that even continuous designs for constrained regions can be appreciably more complicated than those for regular regions.

Here we have 10 unequally replicated support points rather than the six equally replicated points of the simplex lattice design. As we have seen, the calculation of exact designs introduces further complications.   □

***Example* 16.4** Quadratic constraints

The following example is taken from an unpublished report by Z. Iliev. As the result of the first phase of an experiment with a three-component mixture, models have been fitted to two responses. In the second phase measurements are to be made of a third response, but only in that portion of the design region where the other two responses have satisfactory values. The requirements $\hat{Y}_1 \geqslant c_1$ and $\hat{Y}_2 \geqslant c_2$ for specified c_1 and c_2 lead to the quadratic constraints

$$-4.062\ x_1^2 + 2.962\ x_1 + x_2 \geqslant 0.6075$$
$$-1.174\ x_1^2 + 1.057\ x_1 + x_2 \geqslant 0.5019$$

These two quadratics define the smile-shaped experimental region shown in Fig. 16.4. The D-optimum continuous design for the second-order canonical polynomial was found by searching over a fine grid covering this region. The resulting six-point design is given in Table 16.3 and shown in Fig. 16.4. An interesting feature of this design is that, like the D-optimum simplex lattice design, it also has six support points.

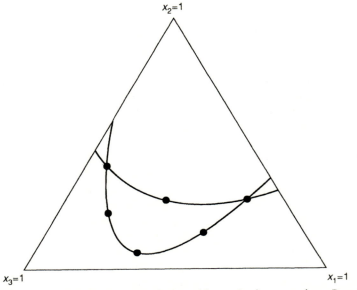

Fig. 16.4. Example 16.4: mixture experiment with quadratic constraints. D-optimum design of Table 16.3.

Table 16.3. Example 16.4: quadratic constraints for a mixture experiment. D-optimum design for the non-regular region of Fig. 16.4

x_1	x_2	x_3
0.59	0.28	0.13
0.51	0.16	0.33
0.34	0.07	0.59
0.33	0.28	0.39
0.16	0.24	0.60
0.07	0.43	0.50

One-sixth of the design weight at each of these points.

This formulation assumes that the constraints defining the design region are known precisely. Allowing for variation in the design region caused by the sampling variability in the coefficients of the estimated constraints is a more complicated design problem, the solution of which requires an extension of the Bayesian methods of Chapter 19.

17
Failure of the experiment and design augmentation

17.1 Failure of an experiment

There are many possible reasons for disappointment or dissatisfaction with the results of an experiment. Four common ones are considered here:

(1) the model is inadequate;

(2) the model does not give reproducible predictions;

(3) many trials failed;

(4) important conditions, often an optimum, lie outside the experimental region.

One cure for several of these experimental shortcomings is to augment the design with some further trials. An example is given in the next section. But first we consider the four possibilities in slightly greater detail.

Inadequacies of the model should be revealed during the course of the kind of statistical analysis outlined in Chapter 8. If the model is inadequate, the investigated relationship will be more complicated than was expected. Systematic departures from the model are often detected by plots of residuals against explanatory variables and by the use of added variable plots. These can suggest the inclusion of higher-order terms in the model, whereas systematic trends in the magnitude of the residuals may suggest the need for a transformation (see, for example, Atkinson 1985). Other patterns in the residuals can sometimes be traced to the effect of omitted or ignored explanatory variables. Examples of the latter, sometimes called 'lurking' variables, are batches of raw materials or reagents, different operators or apparatus, and trends in experimental conditions such as ambient temperature or humidity. These should properly have been included in the experiment as blocking factors or as concomitant observations. Adjustment for these variables after the experiment may be possible, but may cause some loss of balance. In unfortunate cases, when the design becomes far from orthogonal, it may not be possible to distinguish the effects of some factors from those of the omitted variables.

A rather different set of possibilities which may be suggested by the data is that the ranges of some factors are wrong. Excessive changes in the response might suggest a smaller interval for the next part of the experiment, whereas failure to observe an effect for a factor believed to be important would suggest a larger range. Both the revised experimental region that these changes imply and the augmented model consequent on the discovery of specific systematic inadequacies suggest the design of a new experiment. For this the decisions taken at each of the stages in §3.2 should be reconsidered. One possibility is that the resulting new experiment may be an augmentation of the first one, in which case the methods of the next section may be useful.

A rather different situation arises when a model, believed to be adequate, fails correctly to predict the results of new experiments. This may be because of some systematic difference between the new and old observations, for example an unsuspected blocking factor or other lurking variable. Or, particularly for experiments involving many factors, it may be due to the biases introduced in the process of model selection; it is a common occurrence for models to provide much better predictions for the data to which they are fitted than they do on independent sets of data (see, for example, Miller 1990, Chapter 6). A third reason for poor predictions from an apparently satisfactory model is that the experimental design may not permit stringent testing of the assumed model. Considerations of designs for model testing and checking are made in §20.5.

If many individual trials fail there may not be enough data to estimate the parameters of the model. It is natural to think of repeating these missing trials. But it is important to find out if they are missing because of some technical mishap, or whether there is something more fundamentally amiss. Technical mishaps might be accidentally broken or randomly failing apparatus, or a failure of communication in having the correct experiment performed. In such cases the missing trials can be repeated, perhaps with augmentation or modification due to anything that has been learnt from the analysis of the incomplete experiment. On the other hand, the failure may be due to an unsuspected feature of the system being investigated. For example, the failed trials may all lie in a definable subregion of $\mathscr{X}$, in which case the region should be redefined, perhaps by the introduction of constraints. The techniques of Chapter 16 would then be relevant.

Finally, the aim of the experiment may be to define an optimum of the response or of some performance characteristic. If the indication is that this lies far outside the present experimental region, experimental confirmation of this prediction will be required. The strategy of §3.3, together with the design augmentation of the next section, provides methods for moving towards the true optimum.

To entitle this section 'Failure of an experiment' is arguably unduly pessimistic. That an experiment has failed to achieve all the intended goals does not constitute complete failure. Some information will have been obtained which will lead either to abandonment of the project before further resources are wasted or to the planning of a further stage in the experimental programme. As was emphasized in Chapter 3, an experimental study frequently involves many stages which move towards the solution. Inform-ation gathered at one stage should be carefully considered in planning the next stage.

17.2 Design augmentation

We now consider the augmentation of a design by the addition of a specified number of new trials. Examples given in the previous section which lead to augmentation include the need for a higher-order model than can be fitted to the data, different design region, or the introduction of a new factor. In general, we wish to design an experiment incoporating existing data. The new design will then depend on the trials for which the value of the response is known.

Augmentation of a design of size N_0 to one of size N ($N > N_0$) can use any of the criteria described in Chapter 10. When the criterion is D-optimality, the algorithms of Chapter 15 can be employed. However, there are two technical details which need to be attended to. If the design region has changed, the old design points have to be rescaled to the new region using (2.2). These points are then not available for exchange by the algorithm. Second, if the model has been augmented to contain $p > p_0$ parameters, the design for N_0 trials may be singular even for $N_0 > p$. A regularization of this starting design is then required, as described in §15.6. The BLKL algorithm incorporates both these technicalities and so can be used for design augmentation.

Example **17.1**   Augmentation of a second-order design

A second-order model is fitted to the results of a composite design in two factors, i.e. to a 3^2 factorial. For this design and model, $p = 6$ and $N = 9$, so that the model can be tested for adequacy. Suppose that the test shows the model to be inadequate and we would like to obtain a third-order model. One possibility is to start again with a D-optimum design for the third-order model. This will require trials at four values of each x. Figure 17.1 shows a 13-trial D-optimum exact design for the cubic model found by searching over the grid of the 4^2 factorial. Only four of the trials of this design coincide with those of the original design for which the data were collected. Thus nine new trials are indicated. The other possibility is to augment the existing design by searching over the support of the 4^2 factorial. Figure 17.2 shows that the second-order design is augmented by the addition of four points to yield a 13-trial third-order design. In this application of the BLKL algorithm the original nine points are not considered for exchange. The determinant of the information matrix for the augmented design is 0.2236×10^{-7}, which is less than that of the design in Fig. 17.1 for which the value is 0.6601×10^{-7}. Relative to this, the

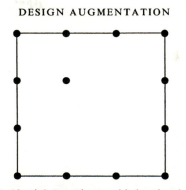

Fig. 17.1. Example 17.1: 13-trial *D*-optimum third-order design found by searching over the points of the 4^2 factorial.

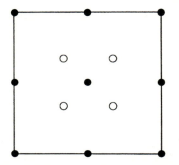

Fig. 17.2. Example 17.1: augmentation of a second-order design to a 13-trial third-order design: ● original second-order design; ○ additional trials.

augmented design, since there are 10 parameters, has a *D*-efficiency of 89.7 per cent. Thus proceeding in two stages using design augmentation has resulted in a loss of efficiency equivalent to the information in just over one trial. The saving is that only four new trials are needed instead of the nine new trials introduced by the design of Fig. 17.1.   □

There are two more general principles raised by this example. The less general is that third-order models are usually not necessary in practice. Transformation of the response, as described in §8.2, is often a more parsimonious elaboration of the model. More general is the principle that augmentation of a design leads to two groups of experiments run at different times, which it may be prudent to treat as coming from different blocks. The augmentation procedure can be extended straightforwardly to handle the blocking variables of Chapter 14. However, in the present example, the design was unchanged by the inclusion of this extra parameter.

18
Non-linear models

18.1 Some examples

Experimental designs for non-linear models depend on the values of the parameters, which will generally be unknown. In this section we first give four examples of non-linear models and then consider the general implications for design of this dependence on unknown values.

Example **18.1** Exponential decay

Many non-linear models arise in the study of chemical kinetics. One of the simplest is the first-order reaction

$$A \to B$$

Measurement is made at time x of the concentration of A. If the initial concentration of A is 1 and the rate constant is θ,

$$\eta(x, \theta) = [A] = \exp(-\theta x). \qquad (x, \theta > 0) \qquad (18.1)$$

Figure 18.1 shows a plot of this response when $\theta = 1$. When $x = 1$, the response $\eta(1, 1) = e^{-1} = 0.368$. So, after one unit, the concentration of A is reduced to 36.8 per cent of the initial value. For larger values of θ the reaction proceeds more rapidly and the response decreases at a greater rate.

This model is closely related to that used in §4.2 to introduce the idea of a non-linear model. If, instead of measuring the concentration of A, that of B is measured, when the initial concentration of A is θ_0, we obtain

$$\eta(x, \theta) = [B] = \theta_0 \{1 - \exp(-\theta_1 x)\},$$

the two-parameter model of §4.2 plotted in Fig. 4.4. In this model initial concentration θ_0 enters linearly. A consequence is that optimum designs will, in general, not depend on the value of θ_0.

Example **18.2** Inverse polynomial

A model with very similar properties to Example 18.1 is the first-order inverse polynomial

$$\eta(x, \theta) = \frac{1}{1 + \theta x} \qquad (x, \theta > 0). \qquad (18.2)$$

At $x = 0$, the response has a value of unity. As x increases, the response decays

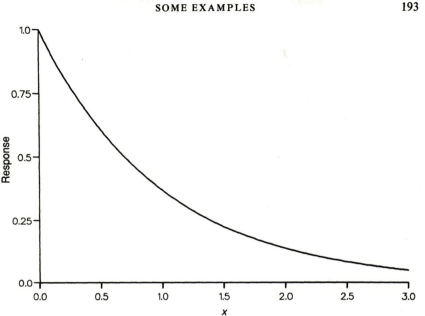

Fig. 18.1. Example 18.1: exponential decay. Response $\eta(x)$.

to zero. For large x, (18.2) decays more slowly than (18.1) However, in practical situations the difference between the models may be hard to detect.

□

Example **18.3** Two consecutive first-order reactions

If the first-order reaction scheme of Example 18.1 is extended to two consecutive first-order reactions

$$A \xrightarrow{\theta_1} B \xrightarrow{\theta_2} C,$$

the concentration of B at time x is

$$\eta(x, \theta) = [B] = \frac{\theta_1}{\theta_1 - \theta_2} \{\exp(-\theta_2 x) - \exp(-\theta_1 x)\} \qquad (x \geqslant 0) \quad (18.3)$$

provided that $\theta_1 > \theta_2 > 0$. Figure 18.2 is a plot of this response for $\theta_1 = 0.7$ and $\theta_2 = 0.2$. Initially, the concentration of B is zero. It rises to a maximum, for these parameter values, at $x = 2.506$, and then decreases more slowly to zero as x increases.

□

Example **18.4** A compartmental model

Example 18.3 is a special case of a class of models which arise frequently in the

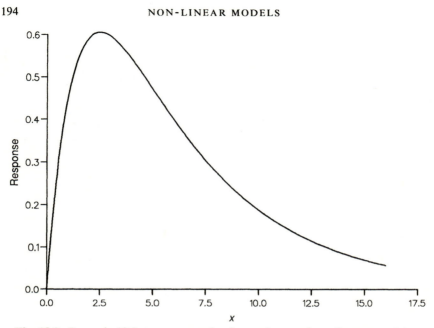

Fig. 18.2. Example 18.3: two consecutive first-order reactions. Response $\eta(x)$.

study of pharmacokinetics, in which the response function is a weighted sum of exponentials of differing exponents. As an example we take

$$\eta(x, \theta) = \theta_3\{\exp(-\theta_1 x) - \exp(-\theta_2 x)\} \qquad (x \geqslant 0) \qquad (18.4)$$

with $\theta_2 > \theta_1$ and all parameters positive. This model, discussed more fully in §18.3, can be thought of as arising from a reaction scheme similar to that yielding (18.3), with the difference that not all of component A is turned into B.

□

A feature common to designs for all four models is that the optimum design will depend on the value of θ. For example, it is clear from Fig. 18.1 that measurements taken when x is near zero, or very large, will not be informative about θ: for a large range of θ values the response will be either close to unity or close to zero at such values of x. Measurements where the expected response is closer to one-half might be expected to be more informative, but the values of x for such response values will depend upon θ.

Because of the dependence of the optimum design on the unknown parameter, sequential designs are more important for non-linear than for linear models. The usual procedure consists of several steps.

1. Start with a preliminary estimate, or guess, of the vector parameter values. This may be either a prior point estimate θ_0 or a prior distribution for θ based on past experience including the analysis of related experiments.

2. The model is linearized by Taylor series expansion.

3. The optimum design is found for the linearized model.

4. A few trials of the optimum design for the linearized model are executed and analysed. If the new estimate of θ is sufficiently accurate the process stops. Otherwise, step 2 is repeated for the new estimate and the process is repeated until sufficient accuracy is obtained.

There are several variants of this scheme. In the most cautious, single trials are added with the model being refitted after each trial. If, as in agricultural field trials, the experiments take a long time to perform, or analyse, such a scheme is clearly unsatisfactory. The opposite extreme to the addition of single trials is, when resources for n experiments are available, to devote about $\sqrt{n}$ of the effort to a preliminary study, with the resulting parameter estimate used in the design of a larger final experiment. In group sequential designs, groups of fixed, or increasing, numbers of trials are added and analysed.

The construction of an optimum sequential scheme depends on the relative costs of experimentation and of time lost in obtaining an accurate answer. In all but the simplest cases the calculations are rarely performed.

The mathematical complexity of even these simple cases, together with an elegant solution, are presented by Gittins (1989). In this chapter we concentrate on the construction of designs for linearized models when the prior information about θ is the point estimate θ_0. The resulting locally optimum designs would be optimum if θ_0 were the true parameter value. Optimum designs for non-linear models when there is a prior distribution for θ are treated in the next chapter.

18.2 *D*-optimum designs

To begin, suppose that θ is a scalar, as in Examples 18.1 or 18.2. Then Taylor expansion of the model yields the linearization

$$E(Y) = \eta(x, \theta) = \eta(x, \theta_0) + (\theta - \theta_0) \frac{\partial \eta}{\partial \theta} (x, \theta)|_{\theta = \theta_0} + \cdots$$

$$= \eta(x, \theta_0) + (\theta - \theta_0) f(x). \tag{18.5}$$

The relationship with linear models is emphasized by rewriting (18.5) as

$$E(Y) - \eta(x, \theta_0) = \beta f(x), \tag{18.6}$$

where $\beta = \theta - \theta_0$. Linearization has thus resulted in a regression model through the origin in which $f(x)$ is the derivative $\partial \eta / \partial \theta$ evaluated at θ_0. The design for which β, and equivalently θ, is estimated with minimum variance is therefore that for which the information matrix $\int f^2(x) \xi(\mathrm{d}x)$ is maximized.

This is achieved by putting all experimental effort at the point where $f(x)$ is a maximum.

Example 18.1 Exponential decay (continued)

For the reponse function (18.1)

$$f(x, \theta) = \frac{\partial \eta}{\partial \theta} = -x \exp(-\theta x), \tag{18.7}$$

which is a function of θ as well as of the single factor x. Evaluation of (18.7) at θ_0 yields the derivative $f(x)$. Figure 18.3 is a plot of $-f(x)$ when $\theta_0 = 1$. This plot starts at zero, rises to a maximum, and then decreases to zero, a shape which makes explicit the idea that experiments at very small or very large values of x are likely to be uninformative.

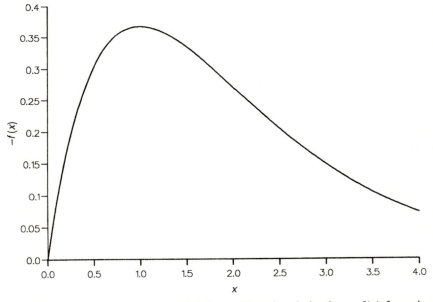

Fig. 18.3. Example 18.1: exponential decay. Negative derivative $-f(x)$ for prior parameter value $\theta_0 = 1$.

The maximum of (18.7) is found by differentiation with respect to x to satisfy the relationship $x\theta = 1$. Thus, to estimate θ with minimum variance, all experiments should be performed when the expected yield is e^{-1}. For the prior value $\theta_0 = 1$, this results in making all the measurements at $x = 1$. For a single parameter model like (18.1) this is the D-optimum design, as well as, of course, the A- and E-optimum one. It shares with many locally D-optimum designs for non-linear models the characteristic of having the number of support points equal to p. Therefore, there are no degrees of freedom available for checking the model. Some approaches to model-checking for non-linear models are mentioned in §20.5, one of which is exemplified in §21.4. Another potential defect is that the design may be inefficient if the value of θ_0 is far from the true θ. The Bayesian optimum designs of the next chapter in which several values of θ are used in the construction of the design provide one way of avoiding overdependence on a single value.

That the design with all trials at $x = 1$ is locally D-optimum can be checked by applying the General Equivalence Theorem, using (18.6). For the linearized one-parameter model, the required variance can be written as

$$d(x,\, \xi,\, \theta) = \frac{f^{2}(x,\, \theta)}{\int f^{2}(x,\, \theta)\xi(dx)} \tag{18.8}$$

For the first-order decay model $f(x, \theta)$ is given by (18.7). Evaluation of (18.8) at $\theta_0 = 1$ for the optimum design which puts all trials at $x = 1$ yields

$$d(x,\, \xi^*) = x^{2}\, \exp\{2(1-x)\}, \tag{18.9}$$

shown plotted in Fig. 18.4. The maximum value $\bar{d}(\xi^*) = p$, for which the maximum of unity occurs at $x = 1$. The design is indeed locally D-optimum.

A last comment on this design is that it has been derived under the assumption common to all designs so far found of additive independent errors of constant variance. In this example, independent errors could arise from repeating the experiment several times, each time taking a single measurement of y at a time x. An alternative form of experiment would be to monitor the evolution of the concentration of A over time, obtaining for each experiment values of y at a series of values of x. The resulting time series of y values would have a correlation structure which should be allowed for in both the design and analysis. Some references on design for time series are given in §22.6.

It may also be that the errors are not of constant variance. In particular, as $\eta \to 0$, the observations will usually be positive. One possibility would be multiplicative errors giving observed values

$$y = \exp(-\theta x)\varepsilon, \tag{18.10}$$

with ε having a non-negative distribution such as the log-normal or the gamma. Taking logarithms on both sides of (18.10) yields the linear model

$$\log y = -\theta x + \log \varepsilon \tag{18.11}$$

If ε is log-normal, so that $\log \varepsilon$ is normal with constant variance, the optimum design for the transformed model (18.11) will take observations at the largest possible value of x, a solution of dubious practical validity which suggests the

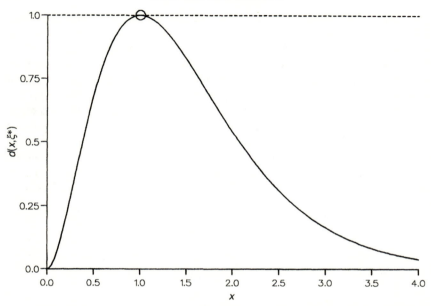

Fig. 18.4. Example 18.1: exponential decay. Variance $d(x, \xi^*)$ for linearized model: ○ local D-optimum design.

inappropriateness in (18.10) of a log-normal ε with constant variance after transformation to (18.11). Designs for non-constant variance are discussed in §22.5   □

The extension of locally D-optimum designs to models with more than one non-linear parameter is straightforward. Let θ be of dimension $p \times 1$, with θ_0 the vector of p prior values of the parameter vector. Then the Taylor expansion which led to (18.6) yields the vector of partial derivatives

$$f^{\mathrm{T}}(x) = \left\{ \frac{\partial \eta(x, \theta)}{\partial \theta_1} \quad \frac{\partial \eta(x, \theta)}{\partial \theta_2} \quad \cdots \quad \frac{\partial \eta(x, \theta)}{\partial \theta_p} \right\} \bigg|_{\theta = \theta_0} \quad (18.12)$$

for the terms of the linearized model.

***Example* 18.3** Two consecutive first-order reactions (continued)

For this two-parameter model, differentiation of (18.3) followed by substitution of the parameter values $\theta_1 = 0.7$ and $\theta_2 = 0.2$ yields the derivatives

$$f_1(x) = (0.8 + 1.4x)\exp(-0.7x) - 0.8\,\exp(-0.2x)$$

$$f_2(x) = (2.8 - 1.4x)\exp(-0.2x) - 2.8\,\exp(-0.7x). \quad (18.13)$$

These are both functions of the single design variable x. The locally D-optimum design as usual maximizes $|M(\xi)|$, where here the rows of the $N \times 2$ extended design matrix F are given by (18.13). The optimum design can be found by use of the methods described in Chapter 9. However, there are some special features which can provide extra insight into the structure of the optimum design so found.

We shall see in a moment that the optimum design has $n = 2$ points of support. As with other D optimum designs with $n = p$, the optimum measure puts weight $1/p$ at each of the supports points, in this case weight $1/2$. For any such design F is a square matrix and

$$|F^{\mathrm{T}}F| = |F|^2. \qquad (18.14)$$

For $p = 1$ the optimum design had all weight concentrated where $f(x)$ was the maximum. The generalization of this result is, for (18.14), that the D-optimum design will maximize $|F|$, which is equivalent to maximizing the volume in the space of the derivatives of the simplex formed by the p points $f(x_i)$ $(i = 1, \ldots, p)$ and the origin.

An example is given in Fig. 18.5, which shows the values of $f_1(x)$ and $f_2(x)$ as a function of x. Experiments at times $x = 0$ or $x = \infty$ are non-informative. For both these values of x, both elements of $f(x)$ are zero and this origin is marked by a circle. As x increases from zero, possible values of $f_1(x)$ and $f_2(x)$ lie on the curve in a clockwise direction. Such a curve is sometimes called a design locus. The optimum design consists of trials at times of 1.23 and 6.86, marked by asterisks in Fig. 18.5, which, together with the origin, form a triangle of maximum area on the design locus.

The maximum yield occurs, as Fig. 18.2 indicates, at $x = 2.506$. The design we have found has one trial at a smaller x value than this and one at a larger. The expected responses at these two points are respectively 83.0 per cent and 56.7 per cent of the maximum. Although the design locus can yield information about the structure of the design, the equivalence theorem provides the confirmation of the optimality of a proposed design. In this case Fig. 18.6 shows the plot of $d(x, \xi^*)$ for the design with support at $x = 1.23$ and $x = 6.86$. For this linearized model $\bar{d}(\xi^*) = 2$, with the maxima at the support points. $\qquad \Box$

18.3 Locally c-optimum designs

The locally D-optimum designs of the previous section are appropriate if interest is in precise estimation of all the parameters of the model. Interest in a subset of the parameters would indicate the use of locally D_s-optimum designs, developed in an analogous manner. In this section we consider designs when estimation of a specified function of the parameters is of interest. We begin with an example.

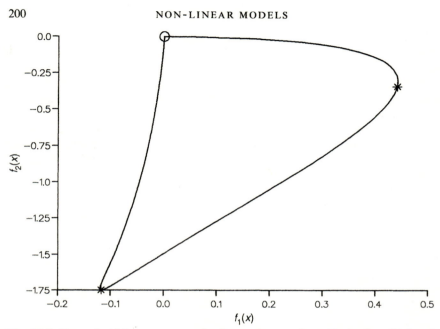

Fig. 18.5. Example 18.3: two consecutive first-order reactions. Derivatives $f_1(x)$ and $f_2(x)$: 'design locus'.

Example **18.4** A compartmental model (continued)

The three-parameter model (18.4) was used by Fresen (1984) to model the results of an experiment in which six horses each received 15 mg/kg of theophylline. Table 18.1 gives the results of 18 measurements of the concentration y of theophylline for one of the horses at a variety of times x. The focus in this example is not whether it is possible, by use of optimum design theory, to do better than this 18-point design. Rather, we shall be concerned with how the optimum design depends upon the aspect of the model that is of interest.

The least squares estimates of the parameters for these results are

$$\theta_1^0 = 0.05884 \qquad \theta_2^0 = 4.298 \qquad \theta_3^0 = 21.80. \qquad (18.15)$$

The fitted curve is similar to Fig. 18.2. There is a rapid rise of the response from zero to the maximum at a time of 1.01 min. Thereafter the exponential decay of the model is governed almost entirely by the value of θ_1, with θ_3 determining the height of the curve.

If the main interest is in estimation of the parameters, the prior estimates (18.15) can be used to find the locally D-optimum design. However, in biological work some derived quantities are often important. We consider three.

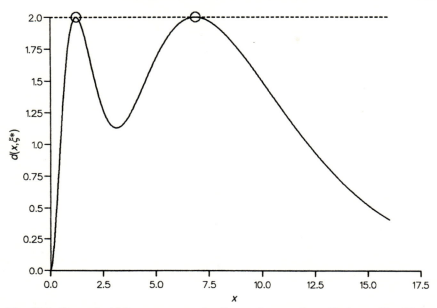

Fig. 18.6. Example 18.3: two consecutive first-order reactions. Variance $d(x, \xi^*)$ for linearized model: ○ local *D*-optimum design.

1. *The area under the curve (AUC)* Model (18.4) is being used rather than (18.3) because not all the theophylline administered appears in the bloodstream; some is metabolized after intragastric administration. The total area under the curve is

$$\int_0^\infty \eta(\theta, x) \, dx = \frac{\theta_3}{\theta_1} - \frac{\theta_3}{\theta_2} = \theta_3\left(\frac{1}{\theta_1} - \frac{1}{\theta_2}\right) = g_1(\theta). \qquad (18.16)$$

Interesting features of this function are that it depends on only two ratios of the parameters and that is linear in θ_3.

2. *Time to maximum concentration* The time to maximum concentration is found by differentiation of $\eta(x, \theta)$ to be

$$x_{\max} = \frac{\log \theta_2 - \log \theta_1}{\theta_2 - \theta_1} = g_2(\theta). \qquad (18.17)$$

For the parameter values of (18.15) $x_{\max} = 1.01$, a value which does not depend on θ_3.

3. *The maximum concentration* The maximum concentration is found by inserting $x_{\max}$ in $\eta(x, \theta)$. The explicit expression for the concentration is complicated without being revealing.   □

Table 18.1. Example 18.4: a compartmental model

Concentration of theophylline in the blood of a horse (μg/ml)

x: time (min)	0.166	0.333	0.5	0.666	1	1.5	2	2.5	3	4	5
y: concentration	10.1	14.8	19.9	22.1	20.8	20.3	19.7	18.9	17.3	16.1	15.0
x (continued)	6	8	10	12	24	30	48				
y (continued)	14.2	13.2	12.3	10.8	6.5	4.6	1.7				

We now consider designs to estimate functions such as (18.16) with minimum variance. In general, we can write the functions as $g(\theta)$. For linear models and linear functions the c-optimum designs of §10.5 minimize

$$\text{var}\{g(\hat{\theta})\} = \text{var}(c^T\hat{\beta}) = c^T\{M(\xi)\}^{-1}c \qquad (18.18)$$

where the $p \times 1$ vector c is of known constants. To estimate a linear combination of the parameters of a non-linear model we would again find designs to minimize (18.18), but the information matrix $M(\xi)$ would be that for the locally linearized model. In order to extend (18.18) to the non-linear function $g(\theta)$ we expand the function in a Taylor series to give the coefficient of θ_i as

$$c_i(\theta) = \frac{\partial g(\theta)}{\partial \theta_i}, \qquad (18.19)$$

evaluated at the prior value θ_0. The resulting locally optimum design will depend on θ both through the coefficients (18.19) which enter the design criterion (18.18) and also through the information matrix $M(\xi)$. To stress the dependence of the designs on θ, the criterion is sometimes called c_θ-optimality

As we shall show, c_θ-optimum designs can have some unexpected properties. For linear models c-optimum designs can be singluar, in which case the equivalence theorem must be adapted slightly (see, for example, Silvey 1980, p. 27). If the design is singular, all parameters of the model will not be estimable, although it must be possible to estimate the linear combination of interest. For non-linear models, or for non-linear functions of the parameters in linear models, the c_θ-optimum design may again be singular, but the property of interest may not be estimable. We first consider some examples, before discussing the implications for design.

Example 18.4 A compartmental model (continued)

Let the three quantities of interest be written as $g_j(\theta)$ $(j=1, 2, 3)$. To obtain linearized combinations of the parameters, let $c_{ij}(\theta) = \partial g_j(\theta)/\partial \theta_i$ $(i=1, 2, 3)$. Then the quantities and coefficients are as given below.

1. The area under the curve:

$$g_1(\theta) = \frac{\theta_3}{\theta_1} - \frac{\theta_3}{\theta_2}$$

$$c_{11}(\theta) = -\frac{\theta_3}{\theta_1^2}$$

$$c_{21}(\theta) = \frac{\theta_3}{\theta_2^2}$$

$$c_{31}(\theta) = \frac{1}{\theta_1} - \frac{1}{\theta_2}.$$

2. The time to maximum concentration:

$$g_2(\theta) = \frac{\log \theta_2 - \log \theta_1}{\theta_2 - \theta_1}.$$

Let $a = \theta_2 - \theta_1$ and $b = \log(\theta_2/\theta_1)$. Then $x_{\max} = b/a$ and

$$c_{12}(\theta) = \frac{b - a/\theta_1}{a^2}$$

$$c_{22}(\theta) = \frac{a/\theta_2 - b}{a^2}$$

$$c_{32}(\theta) = 0.$$

3. The maximum concentration:
$$g_3(\theta) = \eta(x_{\max}, \theta).$$

Let $e_1 = \exp(-\theta_1 x_{\max})$, $e_2 = \exp(-\theta_2 x_{\max})$, and $f = \theta_2 e_2 - \theta_1 e_1$. Then

$$c_{13}(\theta) = \theta_3\{-x_{\max}e_1 + fc_{12}(\theta)\}$$

$$c_{23}(\theta) = \theta_3\{x_{\max}e_2 + fc_{22}(\theta)\}$$

$$c_{33}(\theta) = e_1 - e_2.$$

Table 18.2 shows the locally D- and c-optimum designs. The D-optimum design for this three-parameter model has three support points each with weight 1/3. The plot of the variance $d(x, \xi^*)$ is similar to that of Fig. 18.6 except that there are now three local maxima, each of height 3. This plot introduces no new ideas, and so is not reproduced here.

The D-optimum design, with three support points, permits estimation of the three parameters. The c_θ-optimum designs, on the other hand, with only two or, in one case, one point of support are singular. In order to calculate the design the singularity of $M(\xi)$ was overcome by the use of the ridge-type regularization procedure (10.7) in which a small quantity ε is added to the diagonal of $M(\xi)$ before inversion. An ε value of 10^{-5} was found to be adequate. With this regularization it is possible to check the equivalence theorem that, for each optimum design,

$$\{f(x)^{\mathrm{T}} M^{-1}(\xi^*)c(\theta)\}^2 \leq c^{\mathrm{T}}(\theta)M^{-1}(\xi^*)c(\theta)$$

over the design region $\mathscr{X}$.

The c_θ-optimum design for estimating the area under the curve has only two

Table 18.2. Example 18.4: a compartmental model. D_θ-optimum and c_θ-optimum designs

Criterion	Time x	Design weight	Criterion value
D-optimum	0.2288	1/3	7.3887
	1.3886	1/3	
	18.417	1/3	
AUC	0.2327	0.0135	2194
	17.633	0.9865	
Time to maximum concentration	0.1793	0.6062	0.02815
	3.5671	0.3938	
Maximum concentration	1.0122	1	1.000

points of support. This makes intuitive sense as the criterion is the difference between the ratios θ_3/θ_1 and θ_3/θ_2. The reading at the low x value of 0.23 allows efficient estimation of the ratio θ_3/θ_2 whereas that at $x = 17.6$ is for the ratio θ_3/θ_1. If the response (18.4) is plotted, it is seen that the curve rises very rapidly to the maximum at $x = 1.01$, declining slowly thereafter. The relationship between θ_3 and θ_1 is thus of overwhelming importance in determining the area under the curve, an importance which is reflected in the design's placing almost 99 per cent of the experimental effort at the higher value of x.

The c_θ-optimum design for the time to maximum concentration again has only two points of support. In comparison with the design for the area under the curve, the experimental effort is much more evenly spread over the two design points. The two design points are relatively close to the estimated time of maximum concentration: in this, the design is also very distinct from that of the previous paragraph.

The last of the three functions, the maximum concentration, yields a c_θ-optimum design concentrated on one point: all measurements are taken at x_{max}, the time at which the maximum is believed to occur. This is an example of a c_θ-optimum design for which the quantity of interest is not estimable in the usual sense. If this optimum design were to be used and measurements taken at only one time, it would be impossible to tell where the response was a maximum.

These results demonstrate that, whatever criterion of optimality is used, the optimum design has far fewer points of support than the 18-point design of Table 18.1. As with the one-point design for exponential decay, these optimum designs are derived on the assumption of independent observations, an assumption made in the customary analysis of such data by least squares.

□

Table 18.3. Example 18.4: a compartmental model. Efficiences of D-optimum designs and c-optimum designs of Table 18.2

| Design | Efficiency for | | | |
	D-optimum	AUC	x_{max}	y_{max}
D-optimum	100.0	34.31	66.02	39.10
AUC	0	100	0	0
x_{max}	0	0	100	0
y_{max}	0	0	0	100
18-point	67.61	24.00	28.60	36.77

x_{max}, time to maximum concentration; y_{max}, maximum concentration.

Locally optimum designs for non-linear models incorporate the information in the prior value θ_0. For locally D-optimum designs this information seems not to overwhelm the design criterion: all parameters can be estimated from the design. However, this assertion should perhaps be checked by evaluating the efficiency of the design for a range of parameter values around θ_0. For locally c-optimum designs, the prior value seems to provide too much information, leading to designs which rely so much on the specified value of θ_0 that the properties of interest may not be estimable. There are at least three possible solutions.

1. A general purpose design, such as the locally D-optimum design, can be used even if the purpose is to estimate a specific quantity. Table 18.3 gives the efficiencies of the D_θ- and c-optimum design for Example 18.4. The c_θ-optimum designs all have zero efficiency for quantities other than that for which they are optimum, with the efficiencies being the ratios of the variances (18.18). The efficiency of the D-optimum design varies from 34 to 66 per cent.

2. Several or all quantities of interest can be designed for simultaneously. As the results of Table 18.2 show, the design criteria for the three quantities are of markedly differing magnitudes. Minimization of the sum of the variances would therefore lead to a design in which the largest numerical variance, that for the area under the curve, was given undue prominence. A more satisfactory criterion would be that in which the product of the variances was minimized, so that the criterion to be minimized is

$$\Psi\{M(\xi)\} = \sum_{j=1}^{3} \log\{c_j^{\mathrm{T}}(\theta)M^{-1}(\xi)c_j(\theta)\} \qquad (18.20)$$

a special case of generalized D-optimality (§10.4).

3. If the purpose of the design really is to estimate one quantity with minimum variance, neither of the preceding proposals will provide the best design. The seeming paradox of singular designs arises because of the prior specification of θ_0 as a single value. If θ really is so well known, experimentation is unnecessary. A more realistic specification is to design with a prior distribution for θ which allows for the uncertainty in the parameter values. The resulting Bayesian optimum designs are the subject of the next chapter, with Example 18.4 continued in §19.4.

18.4 The analysis of non-linear experiments

The principles of the analysis of experiments do not depend on whether the models fitted are linear or non-linear. However, there are some details which differ between the two classes of model. In this section a brief introduction is given to fitting non-linear models and to inference about the parameters.

The least squares estimates of the parameters minimize the sum of squares

$$S(\theta) = \sum_{i=1}^{N} \{y_i - \eta(x_i, \theta)\}^2. \qquad (18.21)$$

A straightforward method of calculating the estimates $\hat{\theta}$ is to minimize (18.21) using a standard minimization algorithm such as the NAG routine E04 JAF, a quasi-Newton procedure with numerical calculation of derivatives. However, there are advantages in using the least squares structure of (18.21). One approach is to linearize the model, as in (18.6), and then to proceed iteratively using linear least squares. The procedure can readily be incorporated in such statistical packages as GLIM, with the resulting advantages of flexibility in model specification, plotting of residuals and other quantities, deletion of observations, and ease of calculation of standard errors and test statistics.

More formally, suppose that at stage k of the iterative estimation of θ the parameter estimate is θ^k. Linearization of the p-parameter model yields

$$E(Y) - \eta(x, \theta^k) = (\theta - \theta^k)^{\mathrm{T}} f^k(x) \qquad (18.22)$$

where

$$f^k(x) = \left\{ \frac{\partial \eta(x, \theta)}{\partial \theta_i} \Big|_{\theta=\theta^k} \right\}.$$

For N observations let Z^k be the $N \times 1$ vector with ith element

$$z_i^k = y_i - \eta(x, \theta^k)$$

and let F^k be the extended locally linearized design matrix with ith row $f^{kT}(x_i)$. Application of least squares to (18.22) then yields the estimate

$$\hat{\theta} - \theta^k = (F^{kT}F^k)^{-1}F^{kT}z^k$$

which can be written to provide the Gauss–Newton iteration as

$$\theta^{k+1} = \theta^k + (F^{kT}F^k)^{-1}F^{kT}z^k. \qquad (18.23)$$

The iteration starts from the prior value θ^0 and converges provided that θ^0 is sufficiently close to the least squares estimate $\hat{\theta}$. Often it is not, and (18.23) leads to an increase in the sum of squares, i.e. $S(\theta^{k+1}) > S(\theta^k)$. A useful procedure is then to use (18.23) to define a direction of search for the minimum, but only a move a distance α_k in that direction. The iteration is then to take

$$\theta^{k+1} = \theta^k + \alpha_k(F^{kT}F^k)^{-1}F^{kT}z^k. \qquad (18.24)$$

There are several possibilities for the step length α_k. One is to perform a crude search to find the approximate minimum of $S(\theta)$. Another is to take a fixed value for α_k, provided that $S(\theta)$ does not increase. If it does, the step length is successively reduced until a reduction is achieved in $S(\theta)$.

The problem with increases in $S(\theta)$ when using the Gauss–Newton method can also be avoided by the method of steepest descent. From (18.21)

$$S(\theta^k) = \sum_{i=1}^{N} \{y_i - \eta(x_i, \theta^k)\}^2 = z^{kT}z^k. \qquad (18.25)$$

The gradient direction is, from differentiation of (18.25), $-F^{kT}z^k$. The iteration to decrease the residual sum of squares is then

$$\theta^{k+1} = \theta^k + \alpha_k F^{kT}z^k. \qquad (18.26)$$

where again α_k is a step length. Although (18.26) guarantees a decrease in $S(\theta^k)$ provided that α_k is sufficiently small and the minimum has not been reached, convergence is slow. Near the minimum the Gauss–Newton method is faster. These are combined in the Marquardt–Levenberg iteration

$$\theta^{k+1} = \theta^k + \alpha_k(F^{kT}F^k + \gamma I)^{-1}F^{kT}z^k. \qquad (18.27)$$

When $\gamma = 0$ in (18.27), the Gauss–Newton method is recovered. As $\gamma \to \infty$, steepest descent is obtained. It is usual to scale F^k so that the diagonal elements of $F^{kT}F^k$ are all unity. In use the value of γ is chosen to move towards the Gauss–Newton iteration as the minimum is approached.

The advantage of the Marquardt–Levenberg algorithm is in the flexibility provided by the parameter γ, correct choice of which ensures convergence in a wide variety of problems. A disadvantage is that, unlike the Gauss–Newton algorithm, it does not readily combine with standard statistical packages, so that the advantages of flexibility in data-handling are lost. It is our experience that the Gauss–Newton iteration (18.23) is satisfactory for all but the most intractable problems provided that a crude search is made over values of α_k.

Similar experience for maximum likelihood estimation is reported by Carroll and Ruppert (1988, p. 226), who recommended a first step with $\alpha_k = 1$, with successive halvings of α_k until an improvement is obtained. In our case this is until the residual sum of squares decreases.

Whatever algorithm is used, the output is a point estimate $\hat{\theta}$, yielding the linearized model with extended design matrix $\hat{F}$. An approximate $100(1-\alpha)$ per cent confidence region for θ is given by those values for which

$$(\theta - \hat{\theta})^{\mathrm{T}} \hat{F}^{\mathrm{T}} \hat{F} (\theta - \hat{\theta}) \leqslant ps^2 F_{p,v,\alpha} \qquad (18.28)$$

where $F_{p,v,\alpha}$ is a percentage point of the F distribution, seemingly as in (5.23). However, there are two shortcomings of this approximate theory region.

1. The content of the region is only asymptotically $100(1-\alpha)$ per cent. For practical sized experiments a nominal 95 per cent region might, for example, have a content of 90 or 98 per cent.

2. For linear models the region (18.28) corresponds to a contour of the sum of squares $S(\theta)$. This is not true for most non-linear models, for which the contours of $S(\theta)$ are often asymmetric about $\hat{\theta}$ and may be twisted into the shape of p-dimensional bananas.

References are given in the next section to ways in which improvements can be made to the elliptical contours of (18.28). An advantage of (18.28) is ease of computation. The calculation of contours of $S(\theta)$ is a daunting task, particularly for multi-parameter problems.

18.5 Further reading

Ratkowsky (1990) provides an introduction to non-linear regression models, with an emphasis on the properties of a variety of models for one or a few factors. Locally D-optimum designs for non-linear models were introduced by Box and Lucas (1959), who give a further discussion of Example 18.3. However, they do not mention D-optimality or the General Equivalence Theorem which would confirm the optimality of the designs found. The c_θ-optimum designs of §18.3 are described by Kitsos et al. (1988) and Atkinson et al. (1992) who give more details of Example 18.4.

An encyclopaedic survey of non-linear regression is given by Seber and Wild (1989). Chapter 8 is concerned with compartmental models. Earlier chapters include material on sums of squares contours, confidence regions, and a survey of designs for non-linear models. The survey by Ford *et al.* (1989) is concerned solely with design.

Confidence regions of the form (18.28) arise from first-order expansion of the non-linear model. Methods of obtaining improved regions requiring higher-order information are described by Bates and Watts (1988) and Seber and Wild (1989, Chapter 4). Ratkowsky (1983) provides a comparatively brief introduction. A method for obtaining and presenting intervals of various sizes for a single parameter in a multi-parameter non-linear model is presented by Cook and Weisberg (1990).

19
Optimum Bayesian designs

19.1 Introduction

Optimum designs for non-linear models depend on the values of the vector of unknown parameters θ. In Chapter 18 this dependence was resolved by replacing θ by a prior point estimate θ_0, use of which yielded locally optimum designs. In this chapter we consider instead design when there is a prior distribution for θ, which may be either discrete or continuous. This extra information is incorporated into the design by taking expectations of the design criterion over the prior distribution. A general equivalence theorem for such criteria is given in §19.2. The principal difference between this equivalence theorem and that of §9.2 is that Carathéodory's Theorem no longer provides an upper bound on the number of trials in the experimental design. Thus the designs exhibit the common-sense property that the number of support points increase as the prior for θ becomes more dispersed. This behaviour is illustrated in §19.3 which is concerned with several extensions of D-optimality which incorporate prior information. Section 19.4 describes the parallel extension of c-optimality: as the prior information in this case becomes more dispersed, the singular designs of §18.3 for the compartmental model are replaced by designs with sufficient support points to allow estimation of all parameters. The chapter begins with the three examples which will be used to illustrate all points: two are continued from Chapter 18. In all examples it is assumed that second-order error assumptions hold and that estimation is by least squares.

***Example* 19.1** Truncated quadratic model

The expected value of the response Y is related to the single explanatory variable x by the truncated quadratic relationship

$$E(Y) = \eta(x, \beta, \theta) = \beta x \theta (1 - x\theta) = \beta f(x, \theta) \qquad (0 \leqslant x \leqslant 1/\theta)$$

$$= 0 \qquad \text{otherwise.} \qquad (19.1)$$

For known θ this is a standard linear model with a single parameter β, except that the expected value of the response is constrained to be non-negative. The model is related to those used in pharmacokinetic studies to describe the flow of a drug through a subject, although such models usually involve linear combinations of exponential terms, as does the compartmental model of

Example 19.3. In the pharmacokinetic interpretation $1/\theta$, which would vary between subjects, is the time to complete elimination of the drug. The maximum of the curve, corresponding to the maximum concentration of the drug, is $\beta/4$: interest is in estimation of β. Observations for which $x\theta > 1$ are not informative about the value of β, although they may be about θ.

This simple model demonstrates clearly the properties of designs in the presence of prior information. For a given value of θ the variance of $\hat{\beta}$, the least squares estimate of β, is minimized by putting all trials at the point where the response is maximum, i.e. at $x = 1/2\theta$. If the value of θ is not known a priori, but is described by a prior distribution, this locally optimum design could be used with θ replaced by its expected value. But there may be values of θ within the prior distribution for which concentration of the design on one value of x will give estimates of β with large, or even infinite, variance. The designs derived in this chapter are intended to provide, for example, a small value of the expected variance of $\hat{\beta}$ taken over the distribution of θ. As we shall see, such designs can be very different from those which maximize the expected information about β.   □

Example **19.2** Exponential decay (Example 18.1 continued)

A simple model arising from chemical kinetics, which was discussed in Chapter 18, is that for first-order or exponential decay in which

$$E(Y) = \eta(x, \theta) = \exp(-\theta x) \qquad (x, \theta \geqslant 0). \tag{19.2}$$

Linearization about the prior value θ_0 gave the model

$$E(Y) = \eta(x, \theta_0) + (\theta - \theta_0)f(x), \tag{19.3}$$

where $f(x) = f(x, \theta_0) = -x \exp(-\theta_0 x)$.

The relationship between (19.3) and (19.1) is expressed more forcefully by writing

$$E(Y) - \eta(x, \theta_0) = (\theta - \theta_0)f(x)$$

$$= \beta f(x). \tag{19.4}$$

The variance of $\hat{\beta}$ is again minimized by performing all trials where $f(x)$ is a maximum, i.e. at $x = 1/\theta_0$ when $\eta = e^{-1}$. As in Example 19.1, if the true value of θ is far from θ_0, the variance of $\hat{\beta}$ will be large because experiments will be performed where $f(x)$ is small. In such regions the value of the response is near zero or unity, providing little information about θ.   □

Example **19.3** A compartmental model (Example 18.4 continued)

The properties of the three-parameter model

$$E(Y) = \theta_3\{\exp(-\theta_2 x) - \exp(-\theta_1 x)\} \tag{19.5}$$

were discussed in §18.3 and a variety of designs were derived. The locally D-optimum design found by linearization of the model required three equally replicated design points. The c_θ-optimum designs for estimating the area under the curve and for finding the time to maximum concentration had only two design points, in neither case equally replicated, whereas the design for the maximum concentration required trials at only one value of time. It was argued in §18.3 that use of a single prior value of θ_0 for these designs seemed to provide so much information that singular designs resulted. The Bayesian procedure of §19.4 remedies this defect and leads to non-singular designs from which the properties of interest may be estimated.   □

19.2 A General Equivalence Theorem incorporating prior information

In this section the General Equivalence Theorem of §9.2 is extended to include dependence of the information matrix on a vector parameter θ. For linearized non-linear models dependence on θ is through the vector of p partial derivatives

$$f^T(x, \theta) = \left\{\frac{\partial \eta(x_i, \theta)}{\partial \theta_j}\right\} \qquad j = (1, \ldots, p).$$

For c_θ—optimum designs there is a further dependence through the coefficients $c_{ij}(\theta)$ (equation (18.19)). We write the information matrix as

$$M(\xi, \theta) = \int_{\mathscr{X}} f(x, \theta) f^T(x, \theta) \xi(\mathrm{d}x) = \int_{\mathscr{X}} m(x, \theta) \xi(\mathrm{d}x).$$

The generalization is to consider design criteria of the form

$$\Psi\{M(\xi)\} = E_\theta \Psi\{M(\xi, \theta)\}.$$

For the one-parameter example of §19.1 reasonable extensions of D-optimality would be to find designs to maximize the expected information about the parameter or to minimize the expected variance of the parameter estimate. The results of §19.3.2 show that these designs are not the same.

Similarly, there are several generalizations of D-optimality when θ is a vector. The obvious generalization is to take

$$\Psi_{\mathrm{I}}\{M(\xi)\} = -E_\theta \log|M(\xi, \theta)| = E_\theta \log|M^{-1}(\xi, \theta)|. \tag{19.6}$$

Another possibility is

$$\Psi_{\mathrm{II}}\{M(\xi)\} = \log E_\theta |M^{-1}(\xi, \theta)| \tag{19.7}$$

which, when $p = 1$, reduces to minimizing the expected variance of the parameter estimate. Five possible generalizations of D-optimality are listed in

Table 19.1, together with their derivative functions, for each of which an equivalence theorem holds (Dubov 1971; Fedorov 1981). These criteria are compared in §19.3.2. For the present we notice that, from a Bayesian viewpoint, not all criteria correspond to pre-posterior expected loss, although Criterion I does.

Table 19.1. Equivalence theorem for Bayesian versions of D-optimality: design criteria and derivative functions

Criterion	$\Psi\{M(\xi)\}$	Derivative function $\phi(x, \xi)$
I	$E \log\|M^{-1}\|$	$p - E \operatorname{tr} M^{-1} m(x, \theta)$
II	$\log E\|M^{-1}\|$	$p - E\{\|M^{-1}\| \operatorname{tr} M^{-1} m(x, \theta)\}/E\|M^{-1}\|$
III	$\log\|EM^{-1}\|$	$p - E\{\operatorname{tr} M^{-1}(EM^{-1})M^{-1} m(x, \theta)\}$
IV	$\log\{E\|M\|\}^{-1}$	$p - E\{\|M\| \operatorname{tr} M^{-1} m(x, \theta)\}/E\|M\|$
V	$\log\|EM\|^{-1}$	$p - \operatorname{tr}(EM)^{-1} m(x, \theta)$

EM is short for $E_\theta M(\xi, \theta)$ etc.

These results provide design criteria whereby the uncertainty in the prior estimates of the parameters is translated into a spread of design points. In the standard theory the criteria are defined by matrices $M(\xi)$, which are linear combinations, with positive coefficients, of elementary information matrices $m(x)$ corresponding to designs with one support point. But in the extensions of D-optimality, for example, dependence is on such functions of matrices as $E_\theta M^{-1}(\xi, \theta)$ or $E_\theta |M(\xi, \theta)|$, the non-additive nature of which precludes the use of Carathéodory's Theorem. As a result the number of support points is no longer bounded by $p(p+1)/2$. The examples of the next two sections show how the non-additive nature of the criteria leads to designs with an appreciable spread of the points of support.

19.3 Bayesian D-optimum designs

19.3.1 *The truncated quadratic model*

Example 19.1 (continued)

As a first example of design criteria incorporating prior information we calculate some designs for the truncated model (19.1), concentrating in particular on Criterion II given by (19.7). In this one-parameter example this reduces to minimizing the expected variance of the parameter estimate. We contrast this design with that maximizing the expected information about β.

The derivative function for Criterion II is given in Table 19.1. It is convenient when referring to these derivative to call $d(x, \xi) = p - \phi(x, \xi)$ the expected variance. Then, for Criterion II,

$$d(x, \xi) = \frac{E_\theta\{M^{-1}(\xi, \theta)\}d(x, \xi, \theta)}{E_\theta\{M^{-1}(\xi, \theta)\}}, \qquad (19.8)$$

where $d(x, \xi, \theta) = f^{\mathrm{T}}(x, \theta)M^{-1}(\xi, \theta)f(x, \theta)$. The expected variance is thus a weighted combination of the variance of the predicted response for the various parameter values. In the one-parameter case the weights are the variances of the parameter estimates. It follows from the equivalence theorem that the points of support of the optimum design are at the maxima of (19.8), where $d(x, \xi^*) = p$.

Suppose that the prior for θ is discrete with weight p_m on the value θ_m. The design criterion (19.7) to be minimized is

$$E_\theta M^{-1}(\xi, \theta) = \sum_m \frac{p_m}{f^2(x, \theta_m)}, \qquad (19.9)$$

with $f(x, \theta)$ given in (19.1). To illustrate the properties of the design let the prior for θ put weight 0.2 on the five values 0.3, 0.6, 1, 1.5, and 2. Trials at values of $x > 1/\theta$ yield a zero response. Thus for $\theta = 2$ a reading at any value of x above 0.5 will be non-informative. Unless the design contains some weight at values less than this, the criterion (19.9) will be infinite. Yet the locally optimum designs, at $x = 1/2\theta$, for the three smallest parameter values all concentrate weight on a single x value at or above 0.5.

The expected values required for the criterion (19.9) are found by summing over the five parameter values. Table 19.2 gives three optimum continuous designs for Criterion II. The first design was found by searching over the convex design space [0, 1], and the second and third were found by grid search over 20 and 10 x values respectively. The designs have either two or three points, more than the single point indicated by Carathéodory's Theorem for the locally optimum designs. The design for the coarser grid has three points; the others have two. That the three-point design is optimum can be checked from the plot of $d(x, \xi^*)$ in Fig. 19.1. The expected variance is unity, i.e. p, at the three design points and less than unity at the other seven points of the discrete design region. However, it is 1.027 at $x = 0.35$, which is not part of the coarse grid. Searching over a finer grid leads to the optimum design in which the weights at 0.3 and 0.4 are almost combined, yielding a two-point design. It is clear why the number of design points has changed. But such behaviour is infrequent for the standard design criteria when the additivity property holds. Of course, for such criteria for a single-parameter model, all optimum designs would require only one design point.

The effect of the spread of design points is to ensure that there is no value of θ for which the design is very poor. The appearance of Fig. 19.1 indicates that it

Table 19.2. Example 19.1: truncated quadratic model. Continuous optimum designs ξ^* minimizing the expected variance of the parameter estimate (Criterion II)

	Region			Criterion value
(a) Convex [0, 1]				32.34
x	0.3430	1		
w^*	0.6951	0.3049		
(b) 20-point grid				32.37
x	0.35	1		
w^*	0.7034	0.2966		
(c) 10-point grid				32.95
x	0.3	0.4	1	
w^*	0.4528	0.2406	0.3066	

The design ξ^* puts weight w^* at point x.

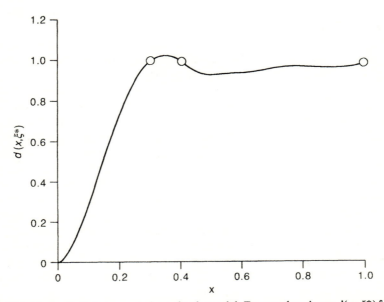

Fig. 19.1. Example 19.1: truncated quadratic model. Expected variance $d(x, \xi^*)$ for the three-point design from searching over the grid $x = 0.1, 0.2, \ldots, 1.0$: $\bigcirc$ design points.

is the sum of several rather different curves arising from the various values of θ. However, not all design criteria lead to a spread of design points. If we use instead a criterion like V of Table 19.1 in which the expected information about β is maximized, (19.9) is replaced by maximization of

$$E_\theta M(\xi, \theta) = \sum_m p_m f^2(x, \theta_m). \qquad (19.10)$$

For the coarse grid the optimum design is at the single point $x = 0.3$. The effect of little or no information about β for a specific θ value may well be outweighed by the information obtained for other θ values. This is not the case for designs using (19.9), when variances can be infinite for some parameter values, whereas the information is bounded at zero.

19.3.2 *A comparison of design criteria*

The results of §19.3.1 illustrate the striking difference between designs which minimize expected variance and those which maximize expected information. In this section we use the exponential decay model (Example 19.2) to compare the five generalizations of *D*-optimality listed in Table 19.1.

When $p = 1$ the five criteria reduce to the three listed in Table 19.3, in which the expectation of integer powers of the information matrix, in this case a scalar, are maximized or minimized as appropriate. The values of the power parameter are also given in Table 19.3. The equivalence theorem for these criteria involves an expected variance of the weighted form

$$d(x, \xi) = \frac{E_\theta\{a(\theta)\, d(x, \xi, \theta)\}}{E_\theta\{a(\theta)\}},$$

where the weights $a(\theta)$ are given in Table 19.3. For Criterion I, $a(\theta) = 1$, so that the combination of variances is unweighted.

Table 19.3. Equivalence theorem for Bayesian versions of *D*-optimality: reduction of criteria of Table 19.1 for single-parameter models

Criterion	$\Psi\{M(\xi, \theta)\}$	Power parameter	Expected variance weight $a(\theta)$
I	$-E_\theta \log M(\xi, \theta)$	0	1
II, III	$E_\theta M^{-1}(\xi, \theta)$	-1	$M^{-1}(\xi, \theta)$
IV, V	$-E_\theta M(\xi, \theta)$	1	$M(\xi, \theta)$

For a numerical comparison of these criteria we use Example 19.2 with, again, five equally probable values of θ, now $1/7$, $1/\sqrt{7}$, 1, $\sqrt{7}$, and 7. For each parameter value the locally *D*-optimum design is at $x = 1/\theta$, so that the design times are uniformly spaced in logarithmic time.

The designs for the three one-parameter criteria are given in Table 19.4. The most satisfactory design arises from Criterion I in which $E_\theta \log|M(\xi, \theta)|$ is maximized. This design puts weights in the approximate ratio of $2:1:1$ within the range of the optimum designs for the individual parameter values. By comparison, the design for Criterion II, in which the expected variance is minimized, puts 96.69 per cent of the weight on $x = 0.1754$. This difference arises because, in the locally D-optimum design for the linearized model, $\text{var}(\hat{\theta}) \propto \theta^2 e^2$. Large parameter values, which result in rapid reactions and experiments at small values of x, are therefore estimated with large variances relative to small parameter values. Designs with Criterion II accordingly tend to choose experimental conditions in order to reduce these large variances. The reverse is true for the design with Criterion V, in which the maximization of expected information leads to a one-point design dominated by the smallest parameter value, for which the optimum design is at $x = 7$: all the weight in the design of Table 19.4 is concentrated on $x = 6.5217$. $\square$

Table 19.4. Example 19.2: exponential decay. Comparison of optimum designs satisfying criteria of Table 19.3

Criterion	Power	x	w^*
I	0	0.2405	0.4781
		1.4863	0.2707
		3.9907	0.2512
II, III	-1	0.1754	0.9669
		2.5529	0.0331
IV, V	1	6.5217	1

The numerical results presented in this section indicate that Criterion I is most satisfactory. We have already mentioned the Bayesian justification for this criterion. A third argument comes from the equivalence theorem. For each value of θ the locally optimum design will have the same maximum value for the variance, in general p. The results of Table 19.3 show that the weight $a(\theta)$ for Criterion I is unity. Therefore the criterion provides an expected variance which precisely reflects the importance of the different θ values as specified by the prior distribution. In other criteria the weights $a(\theta)$ can be considered as distorting the combination of the already correctly scaled variances.

Despite these arguments, there may be occasions when the variance of the parameter estimates is of prime importance and Criterion II is appropriate.

For Example 19.1 this criterion produced an appealing design in §19.3.1 because the variance of $\hat{\beta}$ for the locally optimum design does not depend on θ. But the results of the present section support the use of the Bayesian criterion in which $E_\theta \log|M^{-1}(\xi, \theta)|$ is minimized. In Example 19.2 a further advantage of the design using Criterion I is that a close approximation to the continuous design is found by replacing the weights in Table 19.4 by two, one, and one trials.

19.3.3 *The effect of the prior distribution*

The comparisons of criteria in §19.3.2 used a single five-point prior for θ. In this section the effect of the spread of this prior on the design is investigated together with the effect of more plausible forms of prior. Criterion I is used throughout with Example 19.2.

The more general five-point prior for θ puts weight of 0.2 at the points $1/v$, $1/\sqrt{v}$, 1, $\sqrt{v}$, and v. In §19.3.2 taking $v = 7$ yielded a three-point design. When $v = 1$ the design problem collapses to the locally optimum design with all weight at $x = 1$. Table 19.5 gives optimum designs for these and three other values of v, giving one-, two-, three-, four-, and five-point designs as v increases. The design for $v = 100$ almost consists of weight 0.2 on each of the separate locally optimum designs for the very widely spaced parameter values. A prior with this range but more parameter values might be expected to give a design with more design points. As one example, a nine-point uniform prior with support $v^{-1}, v^{-3/4}, v^{-1/2}, \ldots, v^{3/4}, v$, with v again equal to 100, produces an eight-point design. Rather than explore this path any further, we let Table 19.5 demonstrate one way in which increasing prior uncertainty leads to an increase in the number of design points. In assessing such results, although it may be interesting to observe the change in the designs, it is the efficiencies of the designs for a variety of prior assumptions that is of greater practical importance.

An alternative to these discrete uniform priors in $\log \theta$ is a normal prior in $\log \theta$. This corresponds to a prior assessment of θ values in which $k\theta$ is as likely as θ/k and θ has a log-normal distribution. An effect of continuous priors such as these on the design criteria is to replace the summations in the expectations by integrations. However, numerical routines for the evaluation of integrals reduce once more to the calculation of weighted sums.

The normal distribution used as a prior was chosen to have the same variance τ on the $\log \theta$ scale as the five-point discrete prior with $v = 7$, which gave rise to a three-point design. The normal prior was truncated to have range -2.5τ to 2.5τ, and this range was then divided into seven equal intervals on the $\log \theta$ scale to give weights for the values of θ. To assess the effect of this discretization the calculation was repeated with the prior divided into 15 intervals. The two optimum designs are given in Table 19.6. There are slight

Table 19.5. Example 19.2: exponential decay. Dependence of design on range of prior distribution: optimum designs for Criterion I with five-point prior distribution over $1/v$, $1/\sqrt{v}$, 1, $\sqrt{v}$, and v

v	x	w^*
1	1	1
3	0.6505	0.7690
	1.5750	0.2310
7	0.2405	0.4781
	1.4863	0.2707
	3.9907	0.2512
13	0.1109	0.3371
	0.4013	0.1396
	1.2840	0.1955
	6.1466	0.3279
100	0.0106	0.2137
	0.1061	0.1992
	1.0610	0.2000
	10.6490	0.2009
	99.9987	0.1862

differences between these five-point designs. However, the important results are the efficiencies of Table 19.7, calculated on the assumption that the 15-point normal prior holds. The optimum design for the seven-point prior has an efficiency of 99.95 per cent, indicating the irrelevance of the kind of differences shown in Table 19.6. More importantly, the three-point design for the five-point uniform prior has an efficiency of 92.58 per cent. The four-trial exact design derived from this by replacing the weights in Table 19.4 with two, one, and one trials is scarcely less efficient. The only poor design is the one-point locally optimum design.

19.3.4 *Algorithms and the equivalence theorem*

Results such as those of Table 19.6 suggest that there is appreciable robustness of the designs to mis-specification of the prior distribution. A related interpretation is that the optima of the design criteria are flat for Bayesian

Table 19.6. Example 19.2: exponential decay. Optimum designs for discretized log-normal priors

Prior	x	w^*
7-point	0.1012	0.0873
	0.2299	0.1459
	0.6208	0.3653
	1.6588	0.2671
	4.2274	0.1344
15-point	0.1079	0.1083
	0.3329	0.2489
	0.7415	0.2189
	1.4051	0.2496
	3.7389	0.1743

Table 19.7. Example 19.2: exponential decay. Efficiencies of optimum designs for various priors using Criterion I when the true prior is the 15-point log-normal

Prior used in design	Efficiency %
One-point	23.45
5-point uniform, $v=7$	92.58
Exact design for $v=7$	92.18
7-point log-normal	99.95
15-point log-normal	100

designs. This interpretation is supported by plots of the expected variance for some of the designs of Table 19.6.

The plot of $d(x, \xi^*)$ for the locally optimum design for the exponential decay model, putting all weight at $x=1$, was given in Fig. 18.4. The curve is sharply peaked, indicating that designs with trials far from $x=1$ will be markedly inefficient. However, the curve for the design for the five-point uniform prior with $\mu=7$ (Fig. 19.2) is appreciably flatter, with three shallow peaks at the three design points. The curve for the five-point design for the 15-point normal prior (Fig. 19.3) is sensibly constant over a 100-fold range of x, indicating a very flat optimum.

The flatness of the optima for designs with prior information has positive and negative aspects. The positive aspect, illustrated in Table 19.7, is the near-optimum behaviour of designs quite different from the optimum design; the negative aspect is the numerical problem of finding the precisely optimum design, if such is required.

The standard algorithms of optimum design theory are described in §9.4. They consist of adding weight at the point at which $d(x, \xi)$ is a maximum. For the design of Fig. 18.4, with a sharp maximum, the algorithms converge, albeit relatively slowly, since convergence is first order. For flat derivative functions, such as that of Fig. 19.3, our limited experience is that these algorithms are useless, an opinion supported by the comments of Chaloner and Larntz (1989, §4). One difficulty is that small amounts of weight are added to the design at numerous distinct points; the pattern to which the design is converging does not emerge.

The designs described in this section were found using numerical optimization applied to the design criterion, alternating with inspection of the derivative function to indicate regions in which the search for the optimum should be concentrated. In all examples three iterations of optimization and inspection led to designs in which the maximum expected variance was equal to 1 ± 0.0001, so that the equivalence theorem was sensibly satisfied. The optimization routines used were quasi-Newton algorithms with numerical derivatives: NAG routines on the Cyber at Imperial College and the Vax at the London School of Economics, and CMLIB on the Vax at the University of Minnesota. To apply these algorithms for unconstrained optimization, the transformations described in §9.5 were used to ensure that the design constraints were satisfied.

19.4 Bayesian c-optimum designs

The two examples of §19.3 are both one-parameter non-linear models. In this section Bayesian designs are considered for the three-parameter compartmental model.

Example **19.3** A compartmental model (continued)

Although model (19.5) contains three parameters, θ_3 enters the model linearly and so the value of θ_3 does not affect the D-optimum design. In general, c-optimum designs, even for linear models, can depend on the values of the parameters. However, in this example, the coefficients $c_{ij}(\theta)$ either depend linearly on θ_3 or are independent of it, so that, again, the design does not depend on θ_3. In the calculation of Bayesian optimum designs we can

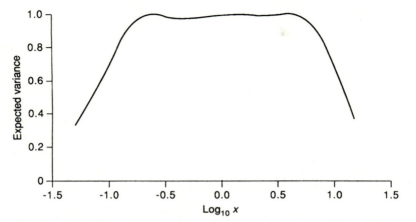

Fig. 19.2. Example 19.2: exponential decay. Expected variance $d(x, \xi^*)$ for Criterion I: five-point uniform prior, $v = 7$.

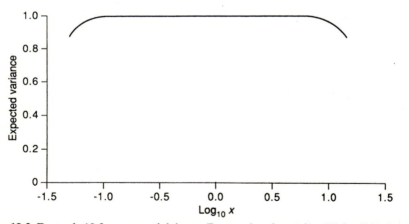

Fig. 19.3. Example 19.2: exponential decay. Expected variance $d(x, \xi^*)$ for Criterion I: 15-point log-normal prior.

therefore take θ_3 to have a fixed value which, as in §18.3, is 21.80. For comparative purposes, two prior distributions are taken for θ_1 and θ_2. These are both symmetric, centred at $(\theta_1, \theta_2) = (0.05884, 4.298)$, the values given in (18.15), and are both uniform over a rectangular region. The calculation of derivatives and the numerical integration required for both c- and D-optimum designs is thus only over these two dimensions of θ_1 and θ_2.

Prior distribution I takes θ_1 to be uniform on 0.05884 ± 0.01 and, independently, θ_2 to be uniform on 4.298 ± 1.0. These intervals are, very approximately, the maximum likelihood estimates for the data of Table 18.1 plus or minus twice the asymptotic standard errors. For prior distribution II the limits are ± 0.04 and ± 4.0, i.e. approximately eight asymptotic standard errors on either side of the maximum likelihood estimator. Prior distribution II thus reprsents appreciably more uncertainty than prior distribution I. Both priors are such that, for all θ_1 and θ_2 in their support, $\theta_2 > \theta_1$, which is a requirement for the model to be of a shape similar to Fig. 18.2.

To find the Bayesian c-optimum designs we minimize $E_\theta c^T(\theta) M^{-1}(\xi, \theta) c(\theta)$, the expected variance of the linear contrasts and the analogue of criterion II of §19.2. An alternative would be the minimization of the expected log variance, which would be the analogue of Criterion I.

For the area under the curve, the Bayesian c-optimum design with prior distribution I is similar to the c_θ-optimum design, but has three design points, not two, as is shown in Table 19.8. However, 95 per cent of the design measure is concentrated at $x = 18.5$. Prior distribution II (Table 19.9) gives an optimum design with four design points, the greatest weight being 70 per cent. These two designs are quite different: increased uncertainty in the parameter values leads to an increased spread of design points. The optimum value of the criterion, the average over the distribution of the asymptotic variance, is also much larger under prior distribution II than under I by a factor of almost 3.

Tables 19.8 and 19.9 also give results for the other two contrasts of interest. For the time to maximum concentration, prior distribution I gives a design with three support points and distribution II gives an optimum design with five. Again, these designs are different from each other and from the c_θ-optimum design. The designs for the maximum concentration also change with the prior information. The c_θ-optimum design has one support point. Prior distribution I gives an optimum design with three points and distribution II gives a design with five points. For prior distribution I the optimum criterion values is only slightly larger than that for c_θ-optimality and for II it is about twice as large. Little is lost by using a prior distribution and estimability is gained.

In addition to the c-optimum designs, Tables 19.8 and 19.9 also include the D-optimum designs found by maximizing $E_\theta \log|M(\xi, \theta)|$, i.e. Criterion I of §19.3. These designs behave much as would be expected from the results of §19.3.3, with the number of support points being three for prior distribution I and five for prior distribution II. The efficiencies of all designs are given in Tables 19.10 and 19.11. The Bayesian c-optimum designs are typically very inefficient for estimation of a property other than that for which they are optimum. This is particularly true under distribution I where, for example, the

Table 19.8. Example 19.3: a compartmental model.
Optimum Bayesian designs for prior distribution I

Criterion	Time x	Design weight	Criterion value
D-optimum	0.2288	0.3333	7.3760
	1.4170	0.3334	
	18.4513	0.3333	
Area under the	0.2449	0.0129	2463.3
curve	1.4950	0.0387	
	18.4903	0.9484	
Time to maximum	0.1829	0.6023	0.030303
concentration	2.4639	0.2979	
	8.8542	0.0998	
Maximum	0.3608	0.0730	1.1144
concentration	1.1446	0.9094	
	20.9218	0.0176	

Bayesian c-optimum design for estimating the area under the curve has an efficiency of about 3 per cent for estimating the time of maximum yield. Both the various D-optimum designs and the original 18-point design are, in contrast, quite robust for a variety of properties. Although it is hard to draw general conclusions from this one example, it is clear that if the area under the curve is of interest, then that should be taken into account at the design stage.

□

19.5 Discussion

The main result of this chapter is the extension of the standard equivalence theorem of §9.2 to incorporate prior information, yielding the General Equivalence Theorem of §19.2. This theorem has then been exemplified by extensions to the familiar criteria of D- and c-optimality. The equivalence theorem for these expectation criteria has a long implicit history. The earliest proof seems to have been due to Whittle (1973), but the implications, particularly for the number of design points, are not clearly stated. The first complete discussion, including examples of designs, is due to Chaloner and Larntz (1989) who consider logistic regression. Chaloner (1988) briefly treats the more general case of design for generalized linear models. Earlier work does not consider either the number of design points, nor the properties of the derivative function, which are of importance in the construction of designs. Läuter (1974, 1976) proves the theorem in the generality required, but only

Table 19.9. Example 19.3: a compartmental model. Optimum Bayesian designs for prior distribution II

Criterion	Time x	Design weight	Criterion value
D-optimum	0.2034	0.2870	7.1059
	1.1967	0.2327	
	2.8323	0.1004	
	5.8229	0.0678	
	20.1899	0.3120	
Area under the curve	0.2909	0.0089	6925.1
	1.7269	0.0365	
	13.0961	0.2570	
	39.5800	0.6976	
Time to maximum concentration	0.2513	0.2914	0.19091
	0.9383	0.2854	
	2.7558	0.1468	
	8.8381	0.2174	
	24.6554	0.0590	
Maximum concentration	0.3696	0.0971	1.9871
	1.1383	0.3584	
	2.4370	0.3169	
	6.0691	0.1634	
	24.0831	0.0641	

Table 19.10. Example 19.3: a compartmental model. Efficiencies of Bayesian D-optimum and c-optimum designs of Table 19.8 under prior distribution I

Design	Efficiency for			
	D-optimum	AUC	x_{max}	y_{max}
D-optimum	100.0	37.0	67.2	39.3
AUC	23.4	100.0	3.2	4.5
x_{max}	57.4	5.1	100.0	19.6
y_{max}	28.2	1.9	12.4	100.0
18-point	68.4	26.0	30.2	41.0

AUC, area under the curve: x_{max}, time to maximum concentration; y_{max}, maximum concentration.

Table 19.11. Example 19.3: a compartmental model.
Efficiencies of Bayesian *D*-optimum and *c*-optimum
designs of Table 19.9 under prior distribution II

Design	Efficiency for			
	D-optimum	AUC	x_{max}	y_{max}
D-optimum	100.0	28.8	64.7	53.7
AUC	23.3	100.0	7.3	10.8
x_{max}	87.6	13.3	100.0	64.3
y_{max}	59.5	10.8	58.2	100.0
18-point	82.9	31.3	77.4	73.8

AUC, area under the curve: x_{max}, time to maximum
concentration; y_{max}, maximum concentration.

gives examples of designs for composite criteria for linear models. Atkinson
and Cox (1974) use the theorem for Criterion I of Table 19.1 with linear
models. Cook and Nachtsheim (1982) are likewise concerned with designs for
linear models. Pronzato and Walter (1985) calculate numerical optimum
designs for some non-linear problems, but do not mention the equivalence
theorem. Fedorov and Atkinson (1988) give a more algebraic discussion of the
properties of the designs for the criteria of Table 19.1. The example of §19.4 is
described in greater detail by Atkinson *et al.* (1992) who also give a more
complete discussion of the independence of the optimum design from the value
of θ_3. For a more general analysis of such independence for *D*- and
D_s-optimum designs, see Khuri (1984). In all applications, if the prior
information used in calculating the designs is also to be used in the analysis of
the experiments, the information matrices used in this chapter require
augmentation by prior information. Pilz (1983, 1989) provides a survey.

A further example of Bayesian optimum design is given in the next chapter,
the subject of which is the design of experiments for discrimination between
regression models. The resulting optimum designs, like those of this chapter,
depend upon the values of unknown parameters. In §20.4 the Bayesian
technique of this chapter is used to define optimum designs maximizing an
expectation criterion. A similar procedure could be followed in §22.5 for
designs for generalized linear models.

20
Discrimination between models

20.1 Two models

So far in this book it has been assumed that the model generating the data is known. Although the model may be quite general, for example a second-order polynomial in m factors, any of the terms in the model may be needed to explain the data. Experiments have therefore been designed to make possible the estimation of all the parameters of interest. In this chapter we consider instead design when a choice is to be made between two or more competing models neither of which is, in general, a special case of the other.

Example 20.1 Two models for decay

In Chapter 18 D_θ-optimum designs were found for the exponential decay model

$$\eta_1(x, \theta_1) = \exp(-\theta_1 x) \qquad (x, \theta_1 \geqslant 0). \tag{20.1}$$

It was mentioned that it might be hard to discriminate between this model and the inverse polynomial

$$\eta_2(x, \theta_2) = \frac{1}{1 + \theta_2 x} \qquad (x, \theta_2 \geqslant 0). \tag{20.2}$$

At what values of x should measurements be made in order to determine which of these two one-parameter models better explains the data?   □

The structure of Example 20.1 is common to the problems investigated in this chapter. We assume that one model is true and then design experiments to determine which it is.

It is clear for Example 20.1 that both models cannot be true. However, this is not so for nested models.

Example 20.2 Constant or quadratic model

Suppose that

$$\eta_1(x, \theta_1) = \gamma \tag{20.3}$$

$$\eta_2(x, \theta_2) = \theta_0 + \theta_1 x + \theta_2 x^2. \tag{20.4}$$

Here the constant model (20.3) is a special case of the quadratic (20.4).

Therefore if (20.3) is true, so is the degenerate form of (20.4) with $\theta_1 = \theta_2 = 0$. One way of treating such nested examples is to introduce a constraint on the parameter values, for example to insist that $\theta_1^2 + \theta_2^2 \geqslant \delta > 0$. This ensures that the two models are separate, with the consequence that (20.4) is an interesting alternative to (20.3) for all parameter values: the response is now constrained to be a function of x, rather than being constant, under the alternative model.

□

In this section we describe optimum designs for discriminating between two models. In the next section an algorithm for the sequential generation of the designs, which is important for practical applications, is illustrated. A brief description of methods for more than two competing models is given in §20.3. Whether there are two or more than two competing models, the resulting optimum designs depend upon unknown parameter values. In §20.4 we derive Bayesian designs, similar to those of Chapter 19, in which prior knowledge about parameter values is represented by a distribution rather than by a point estimate. Finally, in §20.5, consideration of discrimination between nested models leads to the practically important topic of checking models for lack of fit.

The optimum design for discriminating between two models will depend upon which model is true and, often, on the values of the parameters of the true model. Without loss of generality let this be the first model and write

$$\eta_t(x) = \eta_1(x, \theta_1). \tag{20.5}$$

A good design for discriminating between the models will then provide a large lack-of-fit sum of squares for the second model. When the second model is fitted to the data, the least squares parameter estimates will depend on the experimental design as well as on both the value of θ_1 and the errors. In the absence of error the parameter estimates are

$$\hat{\theta}_2(\xi) = \arg \min_{\theta_2} \int_{\mathscr{X}} \{\eta_t(x) - \eta_2(x, \theta_2)\}^2 \xi(dx), \tag{20.6}$$

yielding a residual sum of squares

$$\Delta_2(\xi) = \int [\eta_t(x) - \eta_2\{x, \hat{\theta}_2(\xi)\}]^2 \xi(dx). \tag{20.7}$$

For linear models, $N\Delta_2(\xi)/\sigma^2$ is the non-centrality parameter of the χ^2 distribution of the residual sum of squares for the second model. Designs which maximize $\Delta_2(\xi)$ are called T-optimum, to emphasize the connection with testing models; the letters D and M have already been used, as we have seen, for other criteria (Kiefer 1959). The T-optimum design, by maximizing (20.7), provides the most powerful F test for lack of fit of the second model when the first is true. If the models are non-linear in the parameters, the exact F test is replaced by asymptotic results, but we still design to maximize (20.7).

For linear models with extended design matrices F_1 and F_2 and parameter vectors θ_1 and θ_2 the least squares estimates $\hat{\theta}_2$ minimizing (20.6) are

$$\hat{\theta}_2 = (F_2^T F_2)^{-1} F_2^T F_1 \theta_1. \tag{20.8}$$

Provided that the two models do not contain any terms in common, the non-centrality parameter (20.7) for this exact design is

$$\frac{N \Delta_2(\xi_N)}{\sigma^2} = \theta_1^T \{ F_1^T F_1 - F_1^T F_2 (F_2^T F_2)^{-1} F_2^T F_1 \} \theta_1 \tag{20.9}$$

which makes explicit the dependence of $\Delta_2(\xi_N)$ on the parameters θ_1 of the true model, unless θ_1 is a scalar. In that case designs maximizing (20.9) minimize the variance of estimation of θ_1 in the combined model $E(Y) = F_1 \theta_1 + F_2 \theta_2$, a criterion which does not depend on the value of θ_1. If θ_1 is a vector, but the two models contain terms in common, θ_1 is reduced by the omission of the common terms. More detailed discussion of these topics is given in Chapter 21. For the moment we continue with general models which may be linear or non-linear.

The quantity $\Delta_2(\xi)$ is another example of a convex function to which the General Equivalence Theorem applies. To establish notation for the derivative function let the T-optimum design yield the estimate $\theta_2^* = \hat{\theta}_2(\xi^*)$. Then

$$\Delta_2(\xi^*) = \int_{\mathcal{X}} \{ \eta_t(x) - \eta_2(x, \theta_2^*) \}^2 \xi(dx). \tag{20.10}$$

For this design, the squared difference between the true and predicted responses at x is

$$\psi_2(x, \xi^*) = \{ \eta_t(x) - \eta_2(x, \theta_2^*) \}^2, \tag{20.11}$$

where $\psi_2(x, \xi)$ is the difference for any other design. We then have the equivalence of the following conditions:

(1) The T-optimum design ξ^* maximizes $\Delta_2(\xi)$;
(2) $\psi_2(\xi^*) \leqslant \Delta_2(\xi^*)$ for all $x \in \mathcal{X}$;
(3) at the points of the optimum design $\psi_2(\xi^*) = \Delta_2(\xi^*)$;
(4) for any non-optimum design, i.e. one for which $\Delta_2(\xi) < \Delta_2(\xi^*)$,

$$\sup_{x \in \mathcal{X}} \psi_2(x, \xi) > \Delta_2(\xi^*).$$

These results are, in all important respects, the same as those we have used for

D-optimality and lead to similar methods of design construction and verification.

***Example* 20.3** Two linear models

As a first example of a T-optimum design we look at discrimination between the two models

$$\eta_1(x, \theta_1) = \theta_{10} + \theta_{11}e^x + \theta_{12}e^{-x} \tag{20.12}$$

$$\eta_2(x, \theta_2) = \theta_{20} + \theta_{21}x + \theta_{22}x^2 \tag{20.13}$$

for $-1 \leqslant x \leqslant 1$. Both models are linear in three parameters, and so will exactly fit any three-point design. Designs for discriminating between the two models will therefore need at least four points.

As before, we take the first model as true. Then the T-optimum design depends on the values of the parameters θ_{11} and θ_{12}, but not on the value of θ_{10}, since both models contain a constant. We consider only one pair of parameter values, taking the true model as

$$\eta_t(x) = 4.5 - 1.5e^x - 2e^{-x}. \tag{20.14}$$

This function, which has a value of -1.488 at $x = -1$, rises to a maximum of 1.036 at $x = 0.144$ before declining to -0.131 at $x = 1$. It can be well approximated by the quadratic polynomial (20.13). The T-optimum design for discriminating between the two models is found by numerical maximization of $\Delta_2(\xi)$ to be

$$\xi^* = \begin{Bmatrix} -1 & -0.669 & 0.144 & 0.957 \\ 0.253 & 0.428 & 0.247 & 0.072 \end{Bmatrix} \tag{20.15}$$

for which $\Delta_2(\xi^*) = 1.087 \times 10^{-3}$. A strange feature of this design is that half the weight is on the first and third design points and half on the other two.

For the particular parameter values of (20.14) the design is not symmetrical and does not span the experimental region. It contains only four design points, the minimum number for discrimination between these two three-parameter models. As an illustration of the equivalence theorem, $\psi_2(x, \xi^*)$ is plotted in Fig. 20.1 as an function of x. The maximum value of $\psi_2(x, \xi^*)$ is indeed equal to $\Delta_2(\xi^*)$, the maximum occurring at the points of the optimum design. $\square$

***Example* 20.1** (continued) Two models for decay

Both models in the preceding example were linear in their parameters. We now find the T-optimum design for discriminating between two non-linear models, using as an example the two models for decay (20.1) and (20.2).

Let the first model be true with $\theta_1 = 1$, so that

$$\eta_t(x) = e^{-x} \qquad (x \geqslant 0). \tag{20.16}$$

The T-optimum design again maximizes the non-centrality parameter $\Delta_2(\xi)$ (equation (20.7)); the only small complication introduced by the non-linearity of $\eta_2(x, \theta_2)$ is the iterative calculation of the non-linear least squares estimates $\hat{\theta}_2(\xi)$. The iterative numerical maximization of $\Delta_2(\xi)$ thus includes an iterative fit at each function evaluation.

The T-optimum design when $\theta_1 = 1$ is

$$\xi^* = \begin{pmatrix} 0.327 & 3.34 \\ 0.3345 & 0.6655 \end{pmatrix}, \tag{20.17}$$

a two-point design allowing discrimination between these one-parameter models. We have already seen in §18.2 that the locally optimum design for θ_1 when $\theta_1^0 = 1$ puts all trials at $x = 1$. The design given by (20.17) divides the design weight between points on either side of this value. That this is the T-optimum design is shown by the plot of $\psi_2(x, \xi^*)$ in Fig. 20.2, which has two maxima with the value of $\Delta_2(\xi^*) = 1.038 \times 10^{-2}$.   □

Example **20.2** (continued) Constant or quadratic model

In the two preceding examples it is possible to estimate the parameters of both models from the T-optimum design. However, with the nested models of Example 20.2 the situation is more complicated.

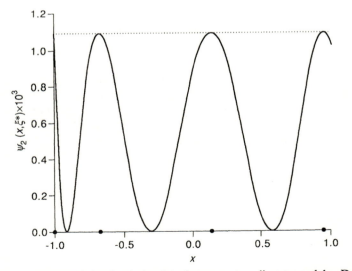

Fig. 20.1. Example 20.3: discrimination between two linear models. Derivative function $\psi_2(x, \xi^*)$ for the T-optimum design: ● design points.

Suppose that the quadratic model (20.4) is true. Then to disprove the constant model (20.3) only requires experiments at two values of x which yield different values of the response. To find such an optimum design let

$$z = \theta_1 x + \theta_2 x^2,$$

the part of model 2 not included in model 1. Then from (20.7) the non-centrality parameter is given by

$$\Delta_1(\xi) = \int_{\mathscr{X}} \{z - \int_{\mathscr{X}} z\xi(dx)\}^2 \xi(dx),$$

since the combined model includes a constant term. The optimum design thus maximizes the sum of squares of z about its mean. This is achieved by a design which places half the trials at the maximum value of z and half at its minimum, i.e. at the maximum and minimum of $\eta_2(x)$. The values of x at which this occurs will depend upon the values of θ_1 and θ_2. But, whatever these parameter values, the design will not allow their estimation.

In order to accommodate the singularity of the design, extensions are necessary to the equivalence theorem defining T-optimality. The details are given by Atkinson and Fedorov (1975a), who provide an analytical derivation

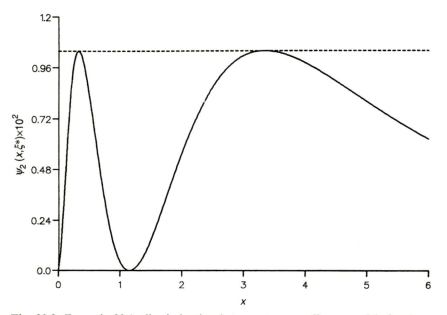

Fig. 20.2. Example 20.1: discrimination between two non-linear models for decay. Derivative function $\psi_2(x, \xi^*)$ for the T-optimum design.

of the T-optimum design for this example. The numerical calculation of singular designs for general problems can be achieved using the regularization (10.7). Atkinson and Fedorov also derive the T-optimum design when the constant model is true, with the alternative the quadratic model constrained so that $\theta_1^2 + \theta_2^2 \geqslant 1$.   □

20.2 Sequential designs for discriminating between two models

The T-optimum designs of the previous section depend both upon which of the two models is true and, in general, on values of the parameters of the true model. They are thus only locally optimum. The Bayesian designs to be described in §20.4 provide one way of designing experiments which are not so dependent on precise, and perhaps limiting, assumptions. In this section we consider an alternative—the sequential construction and analysis of experiments which converge to the T-optimum design, whichever model is true.

The key to the procedure is the estimate of the derivative function $\psi(x, \xi_k)$ after k experiments have been performed and analysed, yielding parameter estimates $\hat{\theta}_{1k} = \hat{\theta}_1(\xi_k)$ and $\hat{\theta}_{2k}$. The corresponding fitted models are then $\eta_1(x, \hat{\theta}_{1k})$ and $\eta_2(x, \hat{\theta}_{2k})$. Since the true model and parameter values are not known, the estimate of the derivative function (20.11) is

$$\psi(x, \xi_k) = \{\eta_1(x, \hat{\theta}_{1k}) - \eta_2(x, \hat{\theta}_{2k})\}^2. \qquad (20.18)$$

In the iterative construction of the T-optimum design using the first-order algorithm (9.17), trials would be added at the point where $\psi_2(x, \xi_k)$ was a maximum. This suggests the following design strategy.

1. After k experiments let the estimated derivative function be $\psi(x, \xi_k)$ given by (20.18).
2. The point x_{k+1} is found for which

$$\psi(x_{k+1}, \xi_k) = \sup_{x \in \mathcal{X}} \psi(x, \xi_k).$$

3. The $(k+1)$th observation is taken at x_{k+1}.
4. Steps 2 and 3 are repeated until sufficient accuracy has been obtained.

Provided that one of the models is true, either $\eta_1(x, \hat{\theta}_{1k})$ or $\eta_2(x, \hat{\theta}_{2k})$ will converge to the true model $\eta_t(x)$ as $k \to \infty$. The sequential design strategy will then converge on the T-optimum design. In order to start the process a design ξ_0, non-singular for both models, is required. In some cases the sequential design converges to a design singular for at least one of the models. In practice this causes no difficulty as ξ_k will be regularized by the starting design ξ_0.

Example **20.3** (continued) Two linear models
The sequential procedure is illustrated by the simulated designs of Fig. 20.3,

using the two linear models with exponential and polynomial terms ((20.12) and (20.13)); the true exponential model is again given by (20.14). In all cases ξ_0 consisted of trials at $-1, 0$, and 1 and, at each stage the sequential design was found by searching over a grid of 21 values of x in steps of 0.1. The efficiency of the sequential design is measured by the ratio of noncentrality parameters $\Delta_2(\xi_k)$ to that of the T-optimum design, where the non-centrality parameter is the residual sum of squares in the absence of error.

Figure 20.3 shows the efficiencies of designs for four increasing values of the error standard deviation σ. For the first design $\sigma = 0$, corresponding to the iterative construction of the optimum design with step-length $\alpha_k = 1/(k+1)$.

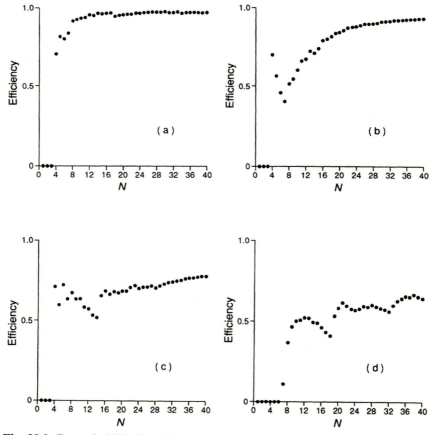

Fig. 20.3. Example 20.3: discrimination between two linear models. Efficiencies of simulated sequential T-optimum designs for increasing error variance: (a) $\sigma = 0$; (b) $\sigma = 0.5$; (c) $\sigma = 1$; (d) $\sigma = 2$.

For increasing value of σ the effect of random fluctuations takes longer and longer to die out of the design. With $\sigma = 0.5$ the design starts to move rapidly towards the optimum after eight trials and, for $\sigma = 1$, it moves to the optimum after 20 trials, whereas, when $\sigma = 2$, the design is still fluctuating after 40 trials. Since for the T-optimum design the maximum difference between the responses is 0.033, even the case of smallest standard deviation ($\sigma = 0.5$) corresponds to an error standard deviation 15 times the effect to be detected. The occasional periods of decreasing efficiency in the plots correspond to the sequential construction of designs for markedly incorrect values of the parameters.   □

It remains only to stress the distinction between the sequential procedure of this section and the iterative algorithms for the construction of designs in §9.4. In the sequential procedure the results of each trial are analysed and parameter estimates are updated before the conditions for the next trial are selected. In the iterative algorithms used in the other examples in this book, the observed response values y_k do not enter. The iterative design procedures depend solely on the values of the factors x.

20.3 Discrimination between more than two models

Suppose, as before, that the first model is true. Then with v models there are $v - 1$ non-centrality parameters

$$\Delta_j(\xi) = \int [\eta_t(x) - \eta_j\{x, \hat{\theta}_j(\xi)\}]^2 \xi(dx) \qquad (20.19)$$

to be maximized, one for each of the false models. Even worse, there are $v(v-1)/2$ functions

$$\psi_{ij}(x, \xi_k) = \{\eta_i(x, \hat{\theta}_{ik}) - \eta_j(x, \hat{\theta}_{jk})\}^2, \qquad (20.20)$$

analogous to (20.18), which could enter into the sequential construction of the design. In order to avoid designs which spend appreciable effort in discriminating between pairs of incorrect models, we find algorithms which take weighted combinations of the squared differences $\psi_{ij}(x, \xi_k)$.

The ideal T-optimum design, if it existed, would simultaneously maximize all non-centrality parameters $\Delta_j(\xi)$. But, almost invariably, a design which is efficient for detecting one kind of departure from the true model does not maximize the non-centrality parameter for some other departure. Therefore we also need weighting in the definition of the T-optimum design as well as in the algorithm. We first consider the properties of the T-optimum design and then derive an algorithm for its sequential construction.

Since the purpose of the experiment is to find the true model, and we are assuming that one model is true, then at some stage, the problem will become that of discriminating between the true model and the model, or models, closest to it. We therefore take as the design criterion the maximization of the non-centrality parameter for the one or more closest models and seek

$$\max_{\xi} \Delta(\xi),$$

where

$$\Delta(\xi) = \min_{j \neq 1} \Delta_j(\xi). \qquad (20.21)$$

We continue to assume that model 1 is true.

In the simplest case when there is only one closest model, this criterion, which corresponds to the choice of one non-zero weight, is a straightforward generalization of the criterion of §20.1. But, in the case studied in this section, maximization of the minimum non-centrality parameter leads to designs for which there are two or more closest models with equal non-centrality parameters. Although there may be two or more models with equal non-centrality parameters, they may not be equally hard to disprove and so may not require equal weights in the design criterion. Let $J^*(\xi)$ be the set of models equally close to model 1 in (20.21), i.e. $J^*(\xi)$ is the solution set of the equation in j

$$\Delta_j(\xi) = \min_{k \neq 1} \Delta_k(\xi). \qquad (20.22)$$

This set will depend upon the design ξ. For all models in $J^*(\xi)$, $\Delta_j(\xi) = \Delta(\xi)$.

To extend the equivalence theorem for T-optimum designs we consider designs with non-zero weights for all members of $J^*(\xi)$. Let the set of non-negative weights be p_j with $\Sigma p_j = 1$, where the summation is over $j \in J^*(\xi)$. Then the T_p-optimum design will maximize the weighted sum of non-centrality parameters

$$\Sigma p_j \Delta_j(\xi) \qquad j \in J^*(\xi). \qquad (20.23)$$

For the optimum design ξ_p^* the equivalence theorem yields the inequality

$$\Sigma p_j \psi_{1j}(x, \xi_p^*) \leq \Sigma p_j \Delta_j(\xi_p^*) \qquad j \in J^*(\xi_p^*). \qquad (20.24)$$

This is the straightforward generalization of the theorem of §20.1 to a linear combination of non-centrality parameters.

For a general set of weights the non-centrality parameters $\Delta_j(\xi_p^*), j \in J^*(\xi_p^*)$ will not be equal. But for one set of weights, which we call p_j^*, the optimum design ξ^* will be such that equality is achieved, so that (20.22) is satisfied for all $j \in J^*(\xi^*)$. This case satisfies the original design criterion (20.21), the equivalence theorem for which yields, in place of (20.24), the inequality

$$\Sigma p_j^* \psi_{1j}(x, \xi^*) \leq \Delta(\xi^*). \qquad (20.25)$$

In order to apply this result we need to calculate both the set of closest models $J^*(\xi^*)$ and the optimum weights p_j^*.

To find a sequential scheme converging to the T-optimum design satisfying (20.25), we begin with the simplest case of only one closest model. The sequential scheme is as follows.

1. After k observations with the design ξ_k the residual sums of squares $S_j(\xi_k)$ are calculated for each model and ranked

$$S_{(1)}(\xi_k) < S_{(2)}(\xi_k) < \ldots < S_{(v)}(\xi_k).$$

For simplicity we assume that a strict ranking is possible.

2. The $(k+1)$th observation is taken at the point x_{k+1} corresponding to

$$\max_{x \in \mathcal{X}} \psi_{(1)(2)}(x, \xi_k),$$

where

$$\psi_{(1)(2)}(x, \xi_k) = \{\eta_{(1)}(x, \hat{\theta}_{(1)k}) - \eta_{(2)}(x, \hat{\theta}_{(2)k})\}^2 \tag{20.26}$$

and (1) and (2) are the indexes of the two models which fit best.

The design given by this sequential procedure converges to the T-optimum design when there is only one closest model. If there is more than one, the situation is more complicated: a procedure is needed which converges to the design satisfying (20.25) incorporating the weights p_j^*. These weights depend upon two or more of the models giving rise to identical values of the non-centrality parameter. However, the path of the sequential design will neither start from such a condition nor will it, in general, pass through it.

One possibility is to apply the simple sequential procedure given above even when there are two or more closest models. If one model is more easily disproved than another, most of the experimental effort will be expended on disproving the former model. Occasionally the sequential procedure will switch to a single trial for the more easily disproved model. This procedure works well for the example at the end of this section. However, an artificial example given by Atkinson and Fedorov (1975b) shows that the simple procedure can converge to a design with a non-centrality parameter appreciably less than that of the T-optimum design.

In order to develop a sequential scheme converging to the T-optimum design satisfying (20.25) we need to accommodate the weights p_j^* and the lack of equality of the non-centrality parameters. For, given $J^*(\xi^*)$ and a fixed set of non-negative weights p_j summing to unity over $J^*(\xi^*)$, the optimum design ξ_p^* could be found by a sequential scheme in which (20.26) is replaced by

$$\max_{x \in \mathcal{X}} \Sigma p_j \psi_{(1)j}(x, \xi_k) \tag{20.27}$$

as the condition yielding x_{k+1}. In (20.27) the summation is over all $j \in J^*(\xi^*)$. From the equivalence theorem (20.25) the optimum weights p_j^* are those for which this maximum is a minimum.

It remains to find the set of closest models $J^*(\xi^*)$. For numerical construction of the optimum design, a method of successive approximation can be used. In this the set $J^*(\xi)$ is replaced by the enlarged set

$$J^{\delta}(\xi) = \{ j: \Delta_j(\xi) - \Delta_1(\xi) \leqslant \delta \} \tag{20.28}$$

the set of models with non-centrality parameters within a tolerance δ of the minimum. For convergence of the iterative algorithm using (20.28) to identify the next design point, δ should presumably decrease with the number of trials k.

There are thus two modifications to be made to the simple sequential procedure (20.26). One is to incorporate the idea of a tolerance into the ranking of the residual sums of squares. The other is to search over the weights minimizing (20.27). The sequential scheme when there are two or more closest models is then as follows.

1. After k observations with the design ξ_k the residual sums of squares $S_j(\xi_k)$ for each model are ranked

    $$S_{(1)}(\xi_k) \leqslant S_{(2)}(\xi_k) \leqslant \ldots \leqslant S_{(v)}(\xi_k).$$

    The set $\hat{J}^{\delta}(\xi_k)$ is found where

    $$\hat{J}^{\delta}(\xi_k) = \{ j: S_j(\xi_k) - S_{(1)}(\xi_k) \leqslant k\,\delta \} \qquad (j \neq j_{(1)}).$$

    Thus $\hat{J}^{\delta}(\xi_k)$ is the set of all models with a residual sum of squares within $k\,\delta$ of the smallest residual sum of squares; the multiplier for sample size k allows for the change from non-centrality parameters for measures in (20.28) to residual sums of squares.

2. The $(k+1)$th observation is taken at x_{k+1} corresponding to

    $$\min_{p_j} \max_{x \in \mathscr{X}} \Sigma p_j \psi_{(1)j}(x, \xi_k). \tag{20.29}$$

The summation in (20.29) is over all $j \in \hat{J}^{\delta}(\xi_k)$. If there is only one closest model, $\hat{J}^{\delta}(\xi_k) = j_{(2)}$ and the procedure reduces to the simple sequential scheme.

Our sequential procedure has therefore led back to maximization of a weighted function of the measures of departure (equation (20.20)). Although the problem is reduced in size by only considering departures from the best-fitting model and giving zero weight to models outside the tolerance δ, the algorithm is appreciably complicated by the two-stage search (20.29).

However, if there are only two closest models, the search over the p_j is one-dimensional, so that the extra complication is not that great.

Example 20.4 Three linear models

In this extension of Example 20.3 there are now three models with $-1 \le x \le 1$:

$$\eta_1(x, \theta_1) = \theta_{10} + \theta_{11}e^x + \theta_{12}e^{-x}$$

$$\eta_2(x, \theta_2) = \theta_{20} + \theta_{21}x + \theta_{22}x^2$$

$$\eta_3(x, \theta_3) = \theta_{30} + \theta_{31}\sin(\tfrac{1}{2}\pi x) + \theta_{32}\cos(\tfrac{1}{2}\pi x) + \theta_{33}\sin(\pi x).$$

As before, we take the first model as true with parameter values given by (20.14). Calculations show that if θ_{33} is put equal to zero, so that both wrong models contain three parameters, we have an example of the simple case in which there is one closest model. The T-optimum design is that for discrimination between models 1 and 2. The reverse situation is obtained by the addition of term in $\cos(\pi x)$ to the trigonometric model, when this augmented form of model 3 is closest. In the form in which the problem is stated here, $\Delta_2(\xi^*) = \Delta_3(\xi^*)$.

To illustrate the properties of sequential designs for these models an initial design was used with trials at -1, -0.7, 0.1, 0.4, and 1. At each succeeding trial the optimum value of x was found by searching over 41 values in steps of 0.05. Figure 20.4 shows the result of the iterative calculation of a 50-trial design equivalent to the simulation of a sequential design in the absence of error, using the simple sequential procedure (20.26) That is, at each trial that experiment is performed which is most efficient for discriminating between the two best-fitting models. Initially, model 2 fits better than model 3, so that trials are added maximizing $\Delta_2(\xi)$. The resulting four-point design is not

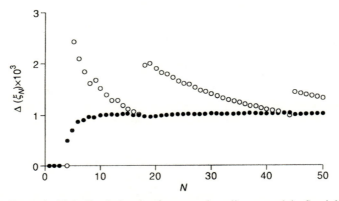

Fig. 20.4. Example 20.4: discrimination between three linear models. Straightforward iterative construction of the design: ● $\Delta_2(\xi_N)$; ○ $\Delta_3(\xi_N)$.

informative about departures from model 3, and $\Delta_3(\xi)$ falls until model 3 fits better. Since model 3 is relatively easily disproved, one trial maximizing $\Delta_3(\xi)$ causes a large increase in the value of $\Delta_3(\xi)$ with a small decrease in $\Delta_2(\xi)$. Trials then proceed accumulating evidence against model 2. Continuation of the simulation shows that this sawtooth pattern in the non-centrality parameters continues with a frequency of about 24 trials, but with decreasing amplitude.

Although this procedure will eventually converge to a stable design, it may not be T-optimum. To illustrate the effect of the tolerance on the sequential scheme, the iterative calculation was repeated with $\delta = 10^{-4}$, about 10 per cent of $\Delta(\xi^*)$. The resulting non-centrality parameters are shown in Fig. 20.5. Initially the design is similar to that of Fig. 20.4. The tolerance first has an effect around trial 20 when, for several trials, model 3 fits better than model 2. Because, with this number of trials, the step-lengths corresponding to the addition of one trial at a time are rather large, the non-centrality parameters soon leave the tolerance, and trial 23, maximizing $\Delta_3(\xi)$, gives another sawtooth jump to the values of the non-centrality parameters. Thereafter the design progresses more smoothly than that of Fig. 20.4.

The effect of experimental error on the procedure is shown in Fig. 20.6, in which pseudo-random normal deviates with $\sigma = 0.05$ were added to the response. The design criterion used the same value of δ as in Fig. 20.5. Because the models are ranked by residual sums of squares in (20.29), for experiments of this size the changeover in design criterion from disproving one model to either disproving the other or a weighted combination of both does not usually occur at or near equal values of the non-centrality parameters.

To illustrate the effect of the tolerance on the properties of the sequential designs, the iterative construction yielding Figs 20.4 and 20.5 was continued

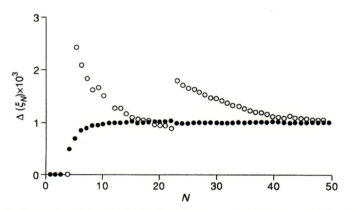

Fig. 20.5. Example 20.4: discrimination between three linear models. Iterative construction of the design with tolerance $\delta = 10^{-4}$: $\bullet$ $\Delta_2(\xi_N)$; $\bigcirc$ $\Delta_3(\xi_N)$.

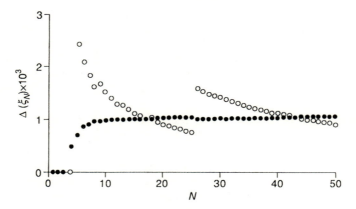

Fig. 20.6. Example 20.4: discrimination between three linear models. Simulated sequential design with $\sigma = 0.05$ and tolerance $\delta = 10^{-4}$, showing the effect of observational error on Fig. 20.5: ● $\Delta_2(\xi_N)$; ○ $\Delta_3(\xi_N)$.

for 250 trials. The results are summarized in Table 20.1 with, for comparison, the sequential designs for the model pairs 1 and 2, and 1 and 3. Since model 3 is relatively easily disproved, the values of $\Delta_2(\xi)$ provide the important comparison. The difference of about 1 per cent in the value of $\Delta_2(\xi)$ between the simple sequential procedure and that using the tolerance is persistent. The equivalence theorem verifies that it is not due to the point at which the construction was stopped.

Although use of the tolerance in this example does not greatly increase $\Delta(\xi)$, Atkinson and Fedorov (1975b) give an example where the design without a tolerance has an efficiency, as measured by the ratio of non-centrality parameters, of only 69 per cent. In the least favourable case with v models, the efficiency could be as little as $1/(v-1)$ if each experiment is informative about departures from only one model. □

20.4 Bayesian designs

The T-optimum designs of the preceding sections depend upon which model is true and on the parameters of the true model. The sequential procedures overcome this defect by converging to the T-optimum design for the true model and parameter values. But, if sequential experiments are not possible, we are left with a design which is only locally optimum. One possibility, as in Chapter 19, is to specify prior distributions and then to take the expectation of the design criterion over this distribution. In this section we revert to the criterion of §20.1 in which there are only two competing models. First, we still assume that we know which model is true, taking expectations only over the

Table 20.1. Example 20.4: three linear models. Comparison of 250 trial sequential designs for discriminating between the models

Design procedure	$\Delta_2(\xi) \times 10^3$	$\Delta_3(\xi) \times 10^3$
First five trials	0.700	2.422
Simple sequential procedure	1.036	1.067
Sequential procedure with tolerance $\delta = 10^{-4}$	1.047	1.082
Sequential procedure for models 1 and 2	1.079	0.088
Sequential procedure for models 1 and 3	0.650	5.668

All sequential designs begin with the same five-trial design.

parameters of the true model. Then we assign a prior probability to the truth of each model and also take expectations of the design criterion over this distribution. In both cases we obtain a straightforward generalization of the equivalence theorem of §20.1.

To begin, we generalize our earlier notation, to make explicit the dependence of the design criterion on the model and parameters. If, as in (20.10), model 1 is true, then the non-centrality parameter is $\Delta_2(\xi, \theta_1)$ with the squared difference in the true and predicted responses (20.11) written as $\psi_2(x, \xi, \theta_1)$. Although we shall not explicitly need the notation, for every design and parameter value θ_1, the least squares estimates of the parameters of the second model (20.6) are $\hat{\theta}_2(\xi, \theta_1)$.

Let E_1 denote expectation with respect to θ_1. Then if we write

$$\Delta_2(\xi) = E_1 \Delta_2(\xi, \theta_1)$$

$$\psi_2(x, \xi) = E_1 \psi_2(x, \xi, \theta_1), \tag{20.30}$$

the equivalence theorem of §20.1 applies to this composite criterion.

Example 20.1 (continued) Two models for decay

The two models (20.1) and (20.2) are respectively exponential decay and an inverse polynomial. In §20.1 we saw that if the exponential model is true with $\theta_1 = 1$, the T-optimum design (20.17) puts design weight at the two points 0.327 and 3.34.

As the simplest illustration of the Bayesian version of T-optimality, we now suppose that the prior distribution of θ_1 assigns a probability of 0.5 to the two values 1/3 and 3, equally spaced from 1 on a logarithmic scale. The T-optimum design is

$$\xi^* = \left\{ \begin{array}{ccc} 0.1160 & 1.073 & 9.345 \\ 0.1608 & 0.4014 & 0.4378 \end{array} \right\} \tag{20.33}$$

with three points of support. For this design the non-centrality parameter $\Delta_2(\xi^*) = 7.056 \times 10^{-3}$. To show that this is indeed the Bayesian T-optimum design, we can again use the equivalence theorem, this time in the form (20.30). Figure 20.7 is a plot of $\psi_2(x, \xi^*)$ against log x. There are three maxima at the design points which are equal to the value of $\Delta_2(\xi^*)$. As in other applications of the theorem we see that the design is optimum.

Other prior distributions give Bayesian designs with more or less points, although always with at least two. However, such calculations provide no new insight into the general properties of the design. Therefore we now consider the design problem when it is not known which of the models is true.   □

Let the prior probability that model j is true be π_j, with, of course, $\pi_1 + \pi_2 = 1$. Then the expected value of the non-centrality parameter, taken over models and over parameters within models is, by extension of (20.30),

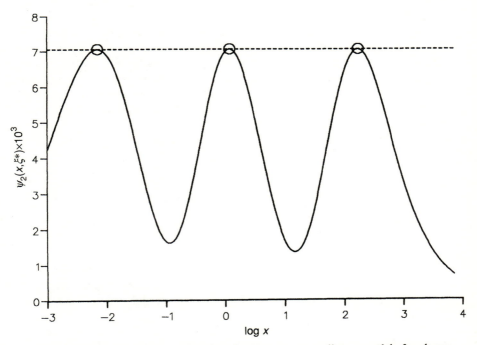

Fig. 20.8. Example 20.1: discrimination between two non-linear models for decay. Derivative function $\psi_2(x, \xi^*)$ for the Bayesian T-optimum design for the two-point prior for θ_1: ○ design points.

$$\Delta(\xi) = \pi_1 E_1 \, \Delta_2(\xi, \theta_1) + \pi_2 E_2 \, \Delta_1(\xi, \theta_2)$$

with the expected squared difference in responses given by

$$\psi(x, \xi) = \pi_1 E_1 \psi_2(x, \xi, \theta_1) + \pi_2 E_2 \psi_1(x, \xi, \theta_2). \tag{20.32}$$

That is, for each model assumed true, we calculate the expected quantity disproving the other model and combine these according to the prior probabilities π_j. The equivalence theorem applies to this more general design criterion as it did to its special case (20.30).

Example 20.3 (continued) Two linear models

The locally T-optimum design for discriminating between these two three-parameter models given in (20.15) put trials at four points. We now consider a prior specification which gives rise to a five-point design.

Table 20.2 gives details of one prior yielding a five-point design. The prior probability that model 1 is true is 0.6 and, conditional on this, there are 10 prior values of the parameters, whereas for model 2 there are only five values. The optimum design

$$\xi^* = \begin{cases} -1.0 & -0.6634 & 0.1624 & 0.8467 & 1 \\ 0.2438 & 0.4265 & 0.2535 & 0.0206 & 0.0556 \end{cases} \tag{20.33}$$

differs from the non-Bayesian design (20.15) in spanning the range of x. However, like that design, it places little experimental effort at the higher values of x.

Table 20.2. Example 20.3: discrimination between two linear models. Prior probabilities of models and prior distribution of parameters yielding (20.31) as the optimum design and the derivative function of Fig. 20.7

$\pi_1 = 0.6$				$\pi_2 = 0.4$			
θ_{10}	θ_{11}	θ_{12}	$p(\theta_1)$	θ_{20}	θ_{21}	θ_{22}	$p(\theta_2)$
4.5	−1.5	−2.0	0.25	1.0	0.5	−2.0	0.23
4.0	−1.0	−2.0	0.14	0.8	0.4	−2.0	0.33
4.5	−2.0	−1.5	0.11	1.0	0.6	−1.5	0.17
5.0	−1.5	−1.5	0.06	1.2	0.5	−1.5	0.15
4.0	−2.0	−1.0	0.05	0.8	0.6	−1.0	0.12
4.5	−1.5	−1.5	0.08				
4.0	−1.5	−2.0	0.05				
4.0	−2.0	−2.0	0.12				
4.5	−2.0	−2.0	0.07				
5.0	−1.5	−2.0	0.07				

The plot of the derivative function $\psi(x, \xi)$ in Fig. 20.8 shows that (20.33) is indeed the optimum design. Comparison of this figure with that for the non-Bayesian design (Fig. 20.1) is informative. In Fig. 20.1, $\psi(x, \xi)$ goes to zero at three points, which is where, for this design, the two response curves intersect. Further experiments at these points would be uninformative about which model is true. However, the corresponding plot for the Bayesian design does not go to zero, as there will always be some combinations of parameter values for which the experiment is informative.   □

These examples illustrate the way in which prior information can be incorporated into designs for discriminating between models. In the second example we assumed independence of the prior distribution between models. It might be more realistic, in some cases, to consider priors which give equal weight to parameter values yielding similarly shaped response curves under the two models. We have also assumed discrete joint prior distributions within models. The case of continuous joint prior distributions would involve no new ideas, but would have to be solved using numerical integration techniques, in themselves a form of discretization. Finally, the extension to three or more models is straightforward when compared with the difficulties encountered in §20.3. There the existence of more than one model closest to the true one led to a design criterion constrained by the equality of non-centrality parameters. Here, expectations can be taken over all $v(v-1)$ non-centrality parameters, yielding a smooth well-behaved design criterion.

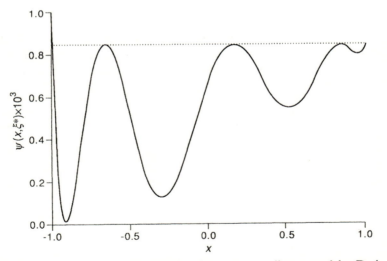

Fig. 20.8. Example 20.3: discrimination between two linear models. Derivative function $\psi(x, \xi^*)$ for the Bayesian T-optimum design for the prior distribution of Table 20.2.

20.5 Checking models and the detection of lack of fit

In this section we shall be particularly concerned with design when one model is a special case of the other. For most of the time we shall assume that the models are linear in the parameters. Some comments on the detection of lack of fit from non-linear models are given at the end of the section.

Suppose that the two models for an N-trial exact design are

$$\eta_1(x, \theta_1) = F_1\theta_1 \tag{20.34}$$

$$\eta_2(x, \theta_2) = F_1\theta_1 + F_2\theta_2 = F\theta. \tag{20.35}$$

Detecting lack of fit from the simpler model (20.34) in the direction (20.35) is equivalent to determing whether θ_2 is zero. The nature of the departure to be detected will determine the terms to be included in F_2. When F_1 is a polynomial of degree d, F_2 usually contains terms of degree $d+1$.

The non-centrality parameter of the F test of the hypothesis $\theta_2 = 0$ is

$$\frac{N \Delta_1(\xi_N)}{\sigma^2} = \theta_2^{\mathrm{T}}\{F_2^{\mathrm{T}}F_2 - F_2^{\mathrm{T}}F_1(F_1^{\mathrm{T}}F_1)^{-1}F_1^{\mathrm{T}}F_2\}\theta_2. \tag{20.36}$$

For discrimination between two non-nested linear models there are two non-centrality parameters, the other being given by (20.9). This subject is pursued further in Chapter 21. The comments following (20.9) apply equally to designs for lack of fit. In particular, if θ_2 in (20.36) is a scalar, the design maximizing (20.36) does not depend on the value of θ_2. If, however, θ_2 is a vector, then the optimum design will depend upon the relative sizes of the elements of θ_2, i.e. upon the particular departure, as well as on its general nature.

There are several possibilities. If sequential design is possible, the sequential T-optimum scheme of §20.2 can be used, as can the Bayesian scheme of §20.4 if there are a number of well-defined departures. Another possibility is the D_s-optimum design for the terms F_2 in the combined model F (20.35), which model contains the terms in both F_1 and F_2. A practical disadvantage of this suggestion is the number of extra terms which have to be estimated. For example, a full second-order model in four factors contains 15 terms. Including all third-order terms adds an extra 20 parameters. The extra experimental effort entailed is particularly unwelcome because, by the nature of model-checking, it is anticipated that all the elements of θ_2 will be zero.

If the departures can be specified as a few linear combinations $C^{\mathrm{T}}\theta$, the D_A-optimum design can be found for these contrasts. However, the problem is often further simplified by the use of designs in which one or more extra trials are added to a good design for the terms in F_1. Two typical examples follow.

Example **20.5** Simple regression

For linear regression in a single factor the expected response is

$$\eta_1(x, \theta_1) = \theta_0 + \theta_1 x.$$

To detect quadratic departures from this model let

$$\eta_2(x, \theta_2) = \theta_0 + \theta_1 x + \theta_2 x^2. \tag{20.37}$$

Then $F_2\theta_2$ in (20.35) is the set of N values of $\theta_2 x_i^2$ ($i = 1, \ldots, N$). The D_s-optimum design for θ_2 when $\mathcal{X} = [-1, 1]$ puts half the trials at $x = 0$ and divides the other half equally between $x = +1$ and $x = -1$.

This design, which leads to the estimation of θ_2 with miniumum variance, has a D-efficiency of $1/\sqrt{2}$ for the first-order model. This may be too great an emphasis on model-checking to be acceptable. In the next chapter we consider multi-criterion designs in which a balance is achieved between powerful model-checking and designs which are efficient for the original model. □

A generalization of Example 20.5 is to take $\eta_1(x, \theta_1)$ to be a polynomial of order $d - 1$ in one factor. In order to check the adequacy of this model the D_s-optimum design is required for the coefficient of x^d when the design region is again $[-1, 1]$. The optimum design (Kiefer and Wolfowitz 1959) was given in §11.4 and has design points

$$x_i = -\cos\left(\frac{i\pi}{d+1}\right) \qquad (i = 0, 1, \ldots, d+1)$$

with weights

$$w_i = \begin{cases} \dfrac{1}{2(d+1)} & (i = 0, d+1) \\[2mm] \dfrac{1}{d+1} & (i = 1, \ldots, d) \end{cases}.$$

Example **20.6** 2^2 factorial plus centre point

In order to detect quadratic departure from simple regression in Example 20.5 extra trials were added at the centre of the design region. The same procedure is often followed for detection of departures from multi-factor first-order models. To be specific assume that

$$\eta_1(x, \theta_1) = \theta_0 + \theta_1 x_1 + \theta_2 x_2 + \theta_{12} x_1 x_2$$

$$\eta_2(x, \theta_2) = \theta_0 + \theta_1 x_1 + \theta_2 x_2 + \theta_{12} x_1 x_2 + \theta_{11} x_1^2 + \theta_{22} x_2^2. \tag{20.38}$$

Let the design be a 2^2 factorial with $x_j = \pm 1$ to which a single centre point is added in order to detect departures from the first-order model. Calculation of the non-centrality parameter (20.36) requires the matrices

$$F_1^T F_1 = \text{diag } (5 \ 4 \ 4 \ 4)$$

$$F_2^T F_2 = \begin{pmatrix} 4 & 4 \\ 4 & 4 \end{pmatrix}$$

$$F_2^T F_1 = \begin{pmatrix} 4 & 0 & 0 & 0 \\ 4 & 0 & 0 & 0 \end{pmatrix}$$

when

$$\frac{5\,\Delta_1(\xi_5)}{\sigma^2} = (\theta_{11}\theta_{22}) \begin{pmatrix} 16/5 & 16/5 \\ 16/5 & 16/5 \end{pmatrix} \begin{pmatrix} \theta_{11} \\ \theta_{22} \end{pmatrix}$$

$$= \frac{16}{5}(\theta_{11} + \theta_{22})^2. \tag{20.39}$$

Thus addition of the single centre point will not lead to the detection of departures for which $\theta_{11} + \theta_{22}$ is small, a result which also folows from comparison of the expected average response at the factorial points and at the centre point. □

In general, trials at the centre of 2^m factorials will fail to detect departures for which $\Sigma\beta_{jj}$ ($j = 1, \ldots, m$) is small. This condition implies either that all β_{jj} are small, or that there is a saddle-point in the response surface. Although uncommon, saddle-points do sometimes occur in the exploration of response surfaces. If a more thorough check is required of a first-order model in m factors, one possibility is the addition to the 2^m factorial not only of a centre point but also of a few extra factorial trials in which one of the x_j is set to zero, preferably a different one in each trial. These are the points for the D-optimum designs for second-order models of §11.5. The power of such designs for detecting lack of fit depends, of course, upon the particular departure which is present. However, the designs do have the advantage of being reasonably efficient for the coefficients of the first-order model.

In order to detect departures from non-linear models we require an analogue to the higher-order polynomial terms used for linear models. Three methods of forming a general model are as follows.

1. Add to the non-linear model a low-order polynomial in m factors. A systematic pattern in the residuals from the non-linear model might be explained by these terms.

2. Add squared and interaction terms in the partial derivatives of the response with respect to the parameters.

3. Embed the non-linear model in a more general model which reduces to the original form for particular values of some non-linear parameters.

These three generalizations are identical for linear models. The first method depends heavily on the design region and is likely to detect departures in those parts of the region which provide little information about the original model. The addition of higher-order terms in the partial derivatives is, on the contrary, invariant under non-linear transformation of the factors. An experimenter with reasonable faith in the model would be advised to use this second method.

20.6 Further reading

Atkinson and Fedorov (1975a) give further details and further examples of designs for discriminating between two models. The connection between these designs and those for parameter estimation is investigated by Fedorov and Khabarov (1986). The material in §20.3 on three or more models is described at greater length by Atkinson and Fedorov (1975b). Ponce de Leon and Atkinson (1991) list additional designs for the Bayesian problem of §20.4 which illustrate the dependence of the design on the prior distribution.

Atkinson and Cox (1974) use generalized D-optimum designs for discriminating between models. Atkinson (1972) discusses the use of related designs for model-checking. These applications are briefly described in Chapter 21. The addition of one or a few trials to a response surface design in order to provide a check on the fitted model is discussed by Shelton *et al.* (1983).

21
Composite design criteria

21.1 Introduction

An experimenter rarely has one purpose or model in mind when designing an experiment. For example, the results of a second-order response surface design will typically be used to test the terms in the model for significance, as well, perhaps, as being used to test for lack of fit. The significant terms will then be included in a model, the estimated response from which may be used for prediction at one or several points not necessarily within $\mathcal{X}$. As we have shown in previous chapters of this book, there are specific design criteria which are addressed to each of these tasks. However, a design which is optimum for one task may be exceptionally inefficient for another. One example is given in Table 11.2, where a comparison is made of several designs in terms of D-efficiency for the first-order model through the origin found appropriate for modelling the desorption of carbon monoxide.

There are two basic approaches which can be followed to provide designs efficient for a variety of purposes. The first is to find a set of near-optimum designs according to a primary criterion and then to choose among them using a secondary criterion. For example, the ACED algorithm of Welch (1984) can provide a set of specified size of the best designs found according to D-optimality, for instance the best 10. Other properties, such as the maximum and average variance of prediction over $\mathcal{X}$, are listed for each of these designs. The final choice between designs is then made by an informal balancing of these criteria.

The second approach is to include all aspects of interest in the design criterion, which is then maximized in the usual way to give either an exact or a continuous design. This approach is explored in the present chapter. The criteria employed are versions of the generalized D-optimality of §10.4. This is first applied to designs for a one-factor polynomial of uncertain degree. The main use of these ideas is for response surface designs in one or more factors where interest may be in both testing and estimation for model-building. In §21.4 these results are extended to non-linear models. The chapter concludes with some remarks on designs for discrimination between models. The designs, unlike those of Chapter 20, are intended for situations in which sequential experimentation is not possible.

21.2 Polynomials in one factor

In §11.4 we discussed D-optimum designs for the one-factor polynomial

$$E(Y) = \beta_0 + \sum_{j=1}^{d} x^j \qquad (21.1)$$

over $\mathscr{X} = [-1, 1]$ when the order d ranged from 2 to 6. The resulting designs had weight $1/(d+1)$ at each of the $d+1$ design points, which were at the roots of an equation involving Legendre polynomials.

It is most unlikely that a sixth-order polynomial would be needed in practice. But it is of interest to see how good the D-optimum sixth-order design is when used for fitting a model of lower order. The D-efficiencies of the sixth-order design relative to the optimum designs of §11.4 are given in Table 21.1. As the order of the polynomial decreases, so does the number of support points of the optimum design. Therefore it is to be expected that the efficiency of the seven-point design for the sixth-order model decreases as the models become simpler. However, the decrease is not very sharp: even for a quadratic the efficiency is 79.7 per cent.

Table 21.1. D-efficiencies of multipurpose D-optimum designs for one-factor polynomials from second to sixth order

	D-efficiency (%)		
Order of model	Sixth-order polynomial	Unweighted composite criterion design (21.2)	Equal-interest design (21.4)
2	79.75	82.76	83.99
3	84.64	88.10	89.11
4	88.74	92.00	92.40
5	92.81	94.95	94.47
6	100	97.26	95.55

We now consider two possible critieria for designs which are efficient for all models from the second order to the sixth. One is to take

$$\Psi\{M(\xi)\} = -\sum_{d=2}^{6} \log|M_d(\xi)|, \qquad (21.2)$$

where $M_d(\xi)$ is the information matrix for the dth-order polynomial (21.1). The dth-order polynomial contains $d+1$ parameters. The D-optimum design for this model is such that, by the equivalence theorem, the maximum variance $\bar{d}_d(\xi^*) = d+1$. For the design criterion (21.2)

$$d(x, \xi) = \sum_{d=2}^{6} d_d(x, \xi)$$

and, for the optimum design,

$$\bar{d}(\xi^*) = \sum_{d=2}^{6} (d+1) = 25. \tag{21.3}$$

This result shows the extent to which the design criterion (21.2) is influenced by models with more parameters. A weighted alternative to (21.2) is the 'equal-interest' design in which each component of the design criterion is standardized by the number of parameters in the component model. Thus (21.2) is replaced by

$$\Psi\{M(\xi)\} = -\sum_{d=2}^{6} \frac{\log|M_d(\xi)|}{d+1} \tag{21.4}$$

with derivative function

$$d(x, \xi) = \sum_{d=2}^{6} \frac{d_d(x, \xi)}{d+1}.$$

For the optimum design, by comparison with (21.3)

$$\bar{d}(\xi^*) = 5. \tag{21.5}$$

In this equal-interest criterion the sum of the D-efficiencies for each model is maximized, as, by the equivalence theorem, is the sum of the G-efficiencies.

The designs maximizing these two criteria, together with the design for the sixth-order model, are given in Table 21.2. The unweighted composite criterion design maximizing (21.2) and the equal-interest design maximizing (21.4) are similar in having more weight at $x = \pm 1$ than the sixth-order design, which has weight $1/7$ at each design point. In every characteristic of design weight and position of design points the composite criterion design lies between the other two. The same is true for the results of Table 21.1 which show the D-efficiencies of the three designs for the individual models from second to sixth order. By moving from the design for the sixth-order model to the equal-interest design, the efficiency for the quadratic is increased from 79.75 to 83.99 per cent with a corresponding decrease in efficiency for the sixth-order model from 100 to 95.55 per cent.

The effect of the criterion on the designs and their properties in Tables 21.1 and 21.2 is slight. A more marked effect is visible in Fig. 21.1 which shows $d(x, \xi)$ for the sixth-order design, which has a maximum value of 7. This

Table 21.2. Multi-purpose *D*-optimum designs for one-factor polynomials from second to sixth order

D-optimum for sixth-order polynomial						
x	−1	0.8302	−0.4688	0	0.4688	0.8302	1
w	0.1429	0.1429	0.1429	0.1429	0.1429	0.1429	0.1429
Unweighted composite criterion design (21.2)						
x	−1	−0.7964	−0.4407	0	0.4407	0.7964	1
w	0.1818	0.1244	0.1286	0.1304	0.1286	0.1244	0.1818
Equal-interest design (21.4)						
x	−1	−0.7891	−0.4346	0	0.4346	0.7891	1
w	0.1954	0.1152	0.1248	0.1292	0.1248	0.1152	0.1954

curve fluctuates appreciably more than that for the equal-interest design with, from (21.5), a maximum value of 5. As would be expected from the Bayesian designs of Chapter 19, averaging the design criterion over five models results in a flatter derivative function at the optimum design.

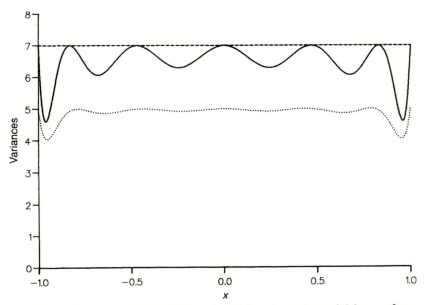

Fig. 21.1. Variances $d(x, \xi^*)$ for designs for sixth-order polynomial in one factor: —— *D*-optimum design for sixth-order polynomial, $\bar{d}(\xi^*) = 7$; ····· equal interest design (21.4), $\bar{d}(\xi^*) = 5$.

The comparisons here are for continuous designs. Exact designs for seven or a few more trials can be expected to show even less effect of changing the design criterion owing to the difficulty of approximating the slight departures of the weights of the design from the uniform distribution for the D-optimum design for the sixth-order model.

21.3 Model-building and parameter estimation

In §20.5 it was suggested that to detect departures from the model

$$E(Y) = F_1\beta_1,\tag{21.6}$$

the model should be extended by the addition of further terms to give a tentative model

$$E(Y) = F_1\beta_1 + F_2\beta_2 = F\beta.\tag{21.7}$$

Experiments should then be designed to estimate β_2, or the non-centrality parameter of the residual sum of squares for (21.6) when (21.7) is true. A disadvantage of this procedure is that the experimental effort is concentrated on determining which model is true, rather than on estimating the parameters of the true model. In this section composite D-optimality is used to provide dual-purpose designs which are efficient for both estimation of the parameters of a model and detection of departures from that model.

Let β_1 be of dimension p and β_2 be of dimension s. We use a criterion which is a weighted combination of the D-optimum design for β_1 and the D_s-optimum design for β_2. For some weight α ($0 \leqslant \alpha \leqslant 1$), the exact design is found which maximizes

$$\frac{\alpha}{p} \log |F_1^T F_1| + \frac{1-\alpha}{s} \log |F_2^T F_2 - F_2^T F_1 (F_1^T F_1)^{-1} F_1^T F_2|$$

$$= \frac{\alpha}{p} \log |F_1^T F_1| + \frac{1-\alpha}{s} \log\left(\frac{|F^T F|}{|F_1^T F_1|}\right).\tag{21.8}$$

When $\alpha = 1$, maximization of (21.8) yields the D-optimum design for β_1, $\alpha = 0$ gives the D_s-optimum design for β_2, and $\alpha = 1/2$ corresponds to the equal-interest design in which equal weight is given to D-efficiency for the two parts of the design criterion.

In order to use the criterion a value of α can be specified which reflects the interest in estimating β_1 relative to checking the model. A preferable alternative is to find the optimum designs for a series of values of α. Calculation of the D-efficiencies of these designs then gives a basis for choice of design and so, implicitly, of α and of a weighting between parameter estimation and model checking.

For continuous designs (21.8) is replaced by the equivalent function of information matrices $M(\xi)$. From the equivalence theorem the derivative function for (21.8) with a continuous design is

$$d(x, \xi) = \frac{\alpha}{p} f_1^{\mathrm{T}}(x) M_{11}^{-1}(\xi) f_1(x) + \frac{1-\alpha}{s} \{ f^{\mathrm{T}}(x) M^{-1}(\xi) f(x) - f_1^{\mathrm{T}}(x) M_{11}^{-1}(\xi) f_1(x) \}.$$

For the optimum design $\bar{d}(\xi^*) = 1$.

As an example we take the estimation of first-order models and the detection of departures from them, starting with the simplest case, that of a single factor.

***Example* 21.1** Linear or quadratic regression?

For simple regression the basic model (21.6) is

$$E(Y_i) = \beta_0 + \beta_1 x_i \tag{21.9}$$

with the augmented model (21.7)

$$E(Y_i) = \beta_0 + \beta_1 x_i + \beta_2 x_i^2. \tag{21.10}$$

As we have repeatedly seen, the D-optimum design for (21.9) over the design region $\mathscr{X} = [-1, 1]$ puts weight 1/2 at $x = \pm 1$. The D_s-optimum design for β_2 puts weight 1/2 at $x = 0$ and equal weight at $x = \pm 1$.

Because of the special structure of these two extreme designs we do not need to maximize (21.8) for specific values of α. Rather, we can study the change in the efficiency of the design as a function of the weight $w/2$ on each end point. For the family of designs

$$\xi = \begin{Bmatrix} -1 & 0 & 1 \\ w/2 & 1-w & w/2 \end{Bmatrix} \qquad (1/2 \leqslant w \leqslant 1) \tag{21.11}$$

the D-efficiency for the two-parameter first-order model is $\sqrt{w}$, whereas the efficiency for the quadratic term on its own is $4w(1-w)$, which is unity when $w = 1/2$. Table 21.3 gives values of these efficiencies for a few values of w. The efficiency for the first-order model is relatively stable, ranging from 70.7 per cent, i.e. $1/\sqrt{2}$, to 100 per cent, whereas the efficiency for detecting departures goes over the whole range from 0 to 100 per cent, corresponding to the absence, or fully weighted presence, of trials at $x = 0$. The exact three-trial design $w = 2/3$, shown in Table 21.3(b), is both small and efficient for both purposes. This is the D-optimum design for the second-order model (21.10). The five-trial design with two trials at $x = \pm 1$, repeated in Table 21.3(b), is slightly more efficient for estimating the first-order model, but appreciably less efficient for detecting second-order departures. $\square$

Table 21.3. *D*-efficiencies of designs for estimating the first-order model (21.9) and for detecting quadratic departures

(a) Continuous designs

	D-efficiencies (%)	
w	First-order model	Quadratic departure
0.5	70.7	100
0.6	77.5	96
0.7	83.7	84
0.8	89.4	64
0.9	94.9	36
1.0	100	0
	Exact designs	
2/3	81.6	88.9
4/5	89.4	64

(b) Exact designs

						D-efficiencies (%)	
x	-1	0	1	N	w	First-order model	Quadratic departure
n_i	1	1	1	3	2/3	81.6	88.9
	2	1	2	5	4/5	89.4	64

The exact designs in Table 21.3(b) were found by calculation of the efficiencies for small designs suggested by the weights of the optimum continous designs. We now extend this method to designs for departures from *m*-factor first-order models.

Example **21.2** Departure from first-order model in *m* factors

When the number of factors *m* is greater than unity the augmented model (21.10) becomes the second-order model

$$E(Y)=\beta_0 + \sum_{j=1}^{m} \beta_j x_j + \sum_{j=1}^{m-1} \sum_{k=j+1}^{m} \beta_{jk} x_j x_k + \sum_{j=1}^{m} \beta_{jj} x_j^2. \qquad (21.12)$$

For model-building and parameter estimation we consider the efficiency of designs according to four criteria.

1. Estimation of the first-order coefficients β_j, for which the D-efficiency of the design is h_1.
2. Estimation of the complete second-order model (21.12), efficiency h_2.
3. Testing for any departures from the first-order model. The optimum design would be the D_s-optimum design for β_{jk} and β_{jj}. The efficiency for detecting these second-order departures is h_2^s.
4. When a 2^m factorial is used, estimation of the interaction terms does not affect efficiency for the first-order model. However, there is a cost associated with being able to estimate the quadratic terms β_{jj}, since extra trials have to be added to the factorial. The efficiency for estimation of the quadratic terms is h_q^s.

We consider only cubical regions and, as in §11.5, restrict attention to designs derived from the 3^m factorial. The sets of points forming this design can be classified according to the number of their non-zero co-ordinates. The composite designs with centre point, factorial points, and star points at the centres of the faces of the cube are all members of the family $(0, 1, m)$ since the star points have one non-zero co-ordinate. Another family, studied by Kôno (1962) and described in §11.5, is, in this notation, $(0, m-1, m)$ which yields continuous D- and D_s-optimum designs for most of the cases of interest. The designs can be described by the weight to to be distributed over the points within each set. Thus, for a design $(0, 1, m)$ with weights (w_0, w_1, w_m) the design weight w_m would be evenly distributed over the 2^m factorial points.

Table 21.4 shows the D- or D_s-optimum continuous designs for $m \leqslant 5$ for the full quadratic model, the second-order terms, and the quadratic terms. The Kôno designs are optimum for most purposes except for the quadratic terms when $m = 4$ or 5. However, the designs for the different purposes do not have the same values of the design weights. The optimum continuous designs for the quadratic terms are given in Table 21.5(b). Unlike the other designs, the designs for the quadratic terms can be realized for an exact number of trials, although the number increases rapidly with m.

In order to find small exact designs we follow Atkinson (1973) and use integer approximations to the $(0, 1, m)$ star-point designs, which in general have fewer support points than the $(0, m-1, m)$ designs. Some exact designs are given in Table 21.5, where n_j is used to represent the number of trials at each design point with j non-zero co-ordinates. The efficiencies are relative to the continuous D-optimum designs of Table 21.4, so that in many cases there is an upper bound on the efficiency of these exact designs, which are not of the Kôno form. A surprising feature of these designs for $m > 2$ is the absence of centre points.

Table 21.4. Continuous D- and D_s-optimum designs for a second-order model over a cubical experimental region

Star-point $(0, 1, m)$ designs

Design for all the terms in the model

Number of factors m	Design weights w_0	w_1	w_m	Efficiency (%) h_2
2	0.096	0.321	0.583	100
3	0	0.345	0.655	98.8
4	0	0.292	0.708	94.0
5	0	0.253	0.747	90.0

Design for second-order terms

m	w_0	w_1	w_m	h_2^s
2	0.176	0.352	0.472	100
3	0	0.429	0.571	99.4
4	0	0.364	0.636	93.6
5	0	0.312	0.688	88.6

Design for quadratic terms

m	w_0	w_1	w_m	h_q^s
2	0.25	0.5	0.25	100
3	0	0.75	0.25	100
4	0	0.8	0.2	92.7
5	0	0.833	0.167	81.1

Kôno $(0, m-1, m)$ designs

Design for all the terms in the model

Number of factors m	Design weights w_0	w_{m-1}	w_m	Efficiency (%) h_2
2	0.096	0.321	0.583	100
3	0.066	0.424	0.510	100
4	0.047	0.502	0.451	100
5	0.036	0.562	0.402	100

Design for second-order terms

m	w_0	w_{m-1}	w_m	h_2^s
2	0.176	0.352	0.472	100
3	0.108	0.475	0.417	100
4	0.072	0.562	0.366	100
5	0.051	0.625	0.324	100

Design for quadratic terms

m	w_0	w_{m-1}	w_m	h_q^s
2	0.25	0.5	0.25	100
3	0.25	0.75	0	100
4	0.2	0.8	0	92.7
5	0.167	0.833	0	81.1

The D- and D_s-efficiencies are defined as follows: h_2, complete second-order model (21.12); h_2^s, the second-order terms in (21.12); h_q^s, only the quadratic terms in (21.12).

Table 21.5 Exact designs for a second-order model over a cubical experimental region

Number of trials at each				Percentage efficiencies for			
Centre point n_0	Star point n_1	Factorial point n_m	Total number of trials N	First-order terms h_1	All terms h_2	Second-order terms h_2^s	Quadratic terms h_q^s
(a) Star-point $(0, 1, m)$ designs							
m = 2							
				Maximum efficiencies →	100	100	100
1	1	1	9	66.7	**97.4**	**98.1**	88.9
2	1	1	10	60.0	94.4	**98.7**	**94.7**
1	1	2	13	76.9	**99.8**	93.5	70.4
4	2	1	16	50.0	83.6	87.6	**100**
6	3	4	34	64.7	97.4	**100.0**	89.0
m = 3							
				Maximum efficiencies →	98.8	99.4	100
0	1	1	14	71.4	**97.6**	**99.4**	75.3
0	4	1	32	50.0	74.5	75.8	100
0	2	3	36	77.8	**98.8**	97.2	61.6
m = 4							
				Maximum efficiencies →	94.0	93.6	92.7
0	1	1	24	75.0	**93.6**	**93.4**	52.2
0	2	1	32	62.5	86.9	89.8	**72.8**

m = 5

n_0	n_1	n_2	n_3	n_4	n_5	N	Maximum efficiencies →			
							h_1	h_2	h_2^s	h_q^s
0	1				$\frac{1}{2}$	26	69.2	90.0	88.6	**81.1**
0	2				$\frac{1}{2}$	36	55.6	**86.8**	**87.5**	48.6
0	1				1	42	81.0	75.9	77.9	**65.8**
0	4				$\frac{1}{2}$	56	42.9	**89.9**	87.2	31.4
0	3				2	94	74.5	60.1	61.3	**77.4**
0	8				$\frac{1}{2}$	96	33.3	89.1	88.6	41.2
								44.3	43.4	81.1

(b) Other designs

m	n_0	n_1	n_2	n_3	n_4	n_5	N	h_1	h_2	h_2^s	h_q^s
3	4	0	1	0	0		16	50.0	74.5	75.8	**100**
4	8	0	0	1	0		40	60.0	83.8	86.5	92.7
	0	4	0	1	0		64	50.0	67.5	67.2	100
5	16	0	0	0	1	0	96	66.7	**88.1**	**90.4**	**81.1**
	16	0	4	0	1	0	256	50.0	62.1	61.0	100

Bold type is used to indicate the most efficient design for a given number of trials except in those cases where the addition of a few more trials causes an appreciable increase in efficiency.

An efficiency of 100 per cent indicates the D-optimum continuous design, whereas 100.0 per cent is used to indicate a very close approximation to the optimum.

For the five-factor designs the value of $\frac{1}{2}$ for n_5 denotes a half-replicate of the full 2^5 factorial.

Table 21.5(b) lists the optimum continuous designs for the second-order terms. These are also exact designs: like the design for $m=1$ with one, two, and one trials at the three design points, these designs also have numbers of trials at each design point which are power of 2. □

The exact designs of Table 21.5 provide a set of relatively small designs which are efficient both for testing parameters for inclusion in a model and for fitting that model. Further designs, for $m \leqslant 7$, are described and tabulated by Pesotchinsky (1975). More efficient designs could be found by maximization of a composite criterion like (21.8) extended to include all subsets of parameters and fitted models that were felt to be of interest. If exact designs were required, the algorithms of Chapter 15 would be appropriate, yielding results parallel to those of Tables 11.6 and 11.8 for D-optimum designs. For N not much greater than the number of parameters in the model, gains in efficiency would also result from moving away from the points of the 3^m factorial, perhaps by using the adjustment algorithm. However, the results of Table 21.5 suggest that the optimum is flat, i.e. small perturbations in the design have little effect on the value of the design criterion. This is particularly the case with multi-criterion designs, where an improvement in one property is often offset by a deterioration in a second, which almost cancel out in the criterion value.

21.4 Non-linear models

The design of experiments for detecting departures from non-linear regression models as well as for estimating the parameters of the model, while similar to the linear case of the previous section, does introduce some new ideas.

***Example* 21.3** First-order growth

Linearization of the first-order growth model

$$E(Y) = 1 - \exp(-\theta t) \qquad (t > \theta) \qquad (21.13)$$

yields

$$f(\theta, t) = \frac{\partial \eta}{\partial \theta} = t \exp(-\theta t)$$

which, apart from a change of sign, is the derivative for the exponential decay model given by (18.7). The value of $f(\theta, t)$ is zero at $t = 0$, rises to a maximum, and then declines again to zero at $t = \infty$. The locally optimum design puts all trials at the maximum value of $f(\theta, t)$, i.e. at $t = 1/\theta_0$, where θ_0 is the prior value of θ.

This one-point design provides no information about departures from the model. Following the prescription of §20.5 a quadratic in the partial derivative can be added, leading to the augmented model

$$E(Y) = \beta_1 f(\theta, t) + \beta_2 f^2(\theta, t).$$

Since f varies between zero and a maximum, the design problem is equivalent to that for the quadratic

$$E(Y) = \beta_1 x + \beta_2 x^2 \qquad (0 \leqslant x \leqslant 1), \tag{21.14}$$

where the corresponding values of t are found as solutions of

$$x = \theta t \exp(1 - \theta t). \tag{21.15}$$

The design for detecting departures from (21.13) is the D_s-optimum design for β_2 in (21.14) which has trials at $\sqrt{2} - 1$ and 1 (9.4). The existence of two values of t satisfying (21.15) for all non-negative $x < 1$ allows increased flexibility in the choice of a design. The resulting optimum design for $\theta = 0.1$ is given in Table 21.6. The times of 1.83 and 29.7 both give the same value of x, and $\sqrt{2}/2$ of the trials can be arbitrarily divided between them without affecting the efficiency of the design for either parameter estimation or the detection of departures.

Table 21.6. Optimum design for detecting departures from the first-order growth model (21.13) by estimation of a quadratic term in the partial derivative $\partial \eta / \partial \theta$

x	time	Weight	h_1 (%)
$\sqrt{2} - 1$	$\begin{cases} 1.83 \\ 29.7 \end{cases}$	0.707	
1	10.0	0.293	41.4

The efficiency of the design of Table 21.6 for estimation of θ is 41.4 per cent. Designs with greater emphasis on parameter estimation can be found by use of the compound design criterion (21.8) with a weight α on parameter estimation. For the augmented non-linear model of this example there are only two parameters and so $p = s = 1$.

Table 21.7 gives optimum continuous designs for parameter estimation and model-checking for a few values of α. Also given are the efficiencies h_1 and h_2^s for estimating θ and for detecting departures respectively. For each design there is a choice of two values of t. In the table the design weight has arbitrarily been equally divided between the two values. Since the design is only locally

Table 21.7. Multi-purpose designs for estimation of the parameter in the first-order growth model (21.13) and for detecting departures from the model

| | | | Efficiencies (%) for | |
| | | | | |
α	Time	Design weight	Parameter estimation h_1	Detection of departures h_2^s
0	1.83	0.354		
	10	0.293	41.4	100
	29.7	0.354		
0.2	2.00	0.321		
	10	0.358	48.5	98.2
	28.6	0.321		
0.4	2.20	0.278		
	10	0.445	57.3	91.2
	27.4	0.278		
0.6	2.45	0.217		
	10	0.566	68.3	76.1
	26.1	0.217		
0.8	2.74	0.131		
	10	0.739	82.3	48.1
	24.8	0.131		
1.0	10	1	100	0

Designs put weight α on parameter estimation.

optimum, the spreading of design points in this way is a sensible precaution against a poor prior value of θ. For example, the design for $\alpha = 0.8$ has reasonably high efficiencies of 82.3 per cent for parameter estimation and 48.1 per cent for detection of departure from the model. It would be well approximated by one trial each at times of 2.74 and 24.8, with six trials at $t = 10$. If this design were too large, three trials at $t = 10$ could be augmented by a single further trial at either $t = 2.74$ or $t = 24.8$.   □

This non-linear example shares with Example 21.1 the feature that a trade-off between efficiencies can be established by inspecting a range of potential designs. One additional feature of the non-linear design is the choice of design conditions from solving (21.15). A second additional feature is that the designs of Tables 21.6 and 21.7 are only locally optimum. If a prior distribution is available for θ, rather than a single value, the Bayesian methods of Chapter 19 can be used to provide a design which is optimum over a range of parameter values.

21.5 Discrimination beween models

As a final application of the properties of composite D-optimum designs we consider again discrimination between models. To begin, we return to the simplest case of discrimination, as in §20.1, but now assume that both models are linear. The general method is to find D_s-optimum designs for the complement of the individual models in a combined model. Such designs will be useful in the absence of prior information about parameters when sequential experiments are not possible.

The two models are

$$E(Y)=F_1\beta_1 \qquad E(Y)=F_2\beta_2, \qquad (21.16)$$

yielding a combined model

$$E(Y)=F_1\beta_1+F_2\beta_2=F\beta. \qquad (21.17)$$

In order to detect departures from $F_1\beta_1$, experiments are designed to estimate β_2. If the two models are separate, i.e. they have no terms in common, $\beta_2=0$ corresponds to model 1 being correct. However, the models will frequently have terms in common, sometimes only a constant, but often other terms as well. The combined model can then be written

$$E(Y)=F\beta=F_1\beta_1+\tilde{F}_2\tilde{\beta}_2=\tilde{F}_1\tilde{\beta}_1+F_2\beta_2 \qquad (21.18)$$

where $\tilde{F}_2\tilde{\beta}_2$ is the complement of $F_1\beta_1$ in the combined model $F\beta$, and similarly for $\tilde{F}_1\tilde{\beta}_1$. Thus, when $F_1\beta_1$ and $F_2\beta_2$ are not separate, the combined model is partitioned into separate models in two different ways. To discriminate between the models we use D_s-optimum designs for $\tilde{\beta}_1$ and $\tilde{\beta}_2$. If the dimensions of $\tilde{\beta}_1$ and $\tilde{\beta}_2$ are $\tilde{s}_1$ and $\tilde{s}_2$, the exact equal-interest design for discrimination maximizes

$$\frac{\log(|F^{\mathrm{T}}F|/|F_2^{\mathrm{T}}F_2|)}{\tilde{s}_1}+\frac{\log(|F^{\mathrm{T}}F|/|F_1^{\mathrm{T}}F_1|)}{\tilde{s}_2} \qquad (21.19)$$

Example 21.4 Two linear models

In §20.1 we found the T-optimum design for discrimination between the two three-parameter models

$$E(Y)=\beta_{10}+\beta_{11}e^x+\beta_{12}e^{-x}$$

$$E(Y)=\beta_{20}+\beta_{21}x+\beta_{22}x^2.$$

The combined model formed from this pair contains only five terms, as both component models contain a constant. The D_s-optimum design for evidence against model 1 maximizes the information about β_{21} and β_{22}, i.e. the coefficients of x and x^2, in the combined model. Similarly, the design for evidence against model 2 is the D_s-optimum design for the coefficients of e^x and e^{-x} in the same model. Since both complements contain two parameters the values of $\tilde{s}_1$ and $\tilde{s}_2$ cancel in the equal-interest criterion (21.19). The continous equal-interest design maximizing (21.19) as a function of information matrices is

$$\zeta^D = \left\{ \begin{matrix} -1 & -0.649 & 0 & 0.649 & 1 \\ 0.140 & 0.259 & 0.202 & 0.259 & 0.140 \end{matrix} \right\}. \tag{21.20}$$

This design has many of the properties already noted for D-optimum designs: there are five parameters, and so there are five support points. The design spans the experimental region and is symmetrical. In all these respects it is different from the four-point T-optimum design (20.15). For the specific set of parameter values yielding (20.15) the efficiency of (21.20) is 60.5 per cent where the efficiency is the ratio of the non-centrality parameters. However, in practice, the T-optimum design will only evolve over a sequence of experiments. Figure 21.2 reproduces the simulation of a sequential T-optimum design from Fig. 20.3(d), adding the efficiency of the equal-interest D-optimum design. In this particular simulation the D-optimum design is more efficient for up to 35 trials. ☐

Equal-interest D-optimum designs will usually have more points of support than the T-optimum designs. Therefore they will be less efficient than the T-optimum design for the true, but unknown, parameter values and model. However, as the above example shows, they may be more efficient in the presence of experimental error, which misleads the sequential generation of the T-optimum design. The D-optimum designs will also be efficient alternatives to the Bayesian T-optimum designs of §20.4 when there is appreciable prior uncertainty.

In deriving the equal-interest criterion (21.19) the only consideration was discrimination between models. The criterion could be extended, as in the previous section, to include D-optimality for estimation of parameters in the models to be fitted. Such an extension involves no new ideas. We turn instead to composite designs for three or more models.

If there are $v \geqslant 3$ models, each may be embedded in a more general model and composite D-optimality then used to obtain an optimum design for detecting departures from each model. Three possible schemes are as follows.

1. Form a general model by combining all the linearly independent terms in the v regression models. To detect departures for model i in the direction of the other models, the D_s-optimum design would be used for the

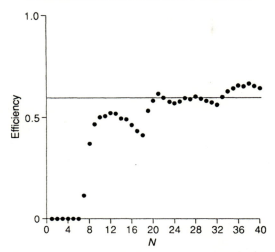

Fig. 21.2. Example 21.1: two linear models. Simulation of a sequential T-optimum design for discrimination between two models: —— efficiency (60.5 per cent) of the equal-interest D-optimum design for discrimination.

complement of the ith model in the combined model. This is the simplest generalization of (21.19), the criterion for two models.

2. Form v general models of the type described in §20.5, each specific for departures from one of the models.

3. Form the $v(v-1)/2$ combined models from all pairs of possible models. As in 1 and 2 a composite D-optimum criterion can be used, with appropriate weightings for parameter estimation and model discrimination.

Table 21.8. Three rival models with additional terms for the detection of departures

Model	Additional term	Efficiency for parameter estimation	Efficiency for detection of departures
First-order growth $\eta_1 = 1 - e^{-\theta t}$	$(\partial \eta_1/\partial \theta)^2$	h_1	h_1^c
Inverse polynomial $\eta_2 = t/(\phi + t)$	$(\partial \eta_2/\partial \phi)^2$	h_2	h_2^c
Quadratic $\eta_3 = \beta_1 t + \beta_2 t^2$	t^3	h_3	h_3^c

Table 21.9 Properties of equal-interest D-optimum designs for discriminating between three rival models

Criterion	Time	Weight	Efficiencies (%) for parameter estimation			Efficiencies (%) for detection of departures		
			h_1	h_2	h_3	h_1^c	h_2^c	h_3^c
Cubic departures from quadratic	5.88	0.455						
	22.0	0.366	55.0	66.9	59.5	35.1	68.3	100
	30.0	0.179						
Equal interest in complements of the three models	1.88	0.327						
	8.49	0.214	41.7	55.2	55.5	76.1	40.6	66.8
	21.4	0.244						
	30.0	0.216						
Equal interest in departures from individual models	7.57	0.351						
	24.0	0.338	49.7	56.5	71.7	74.2	70.8	85.0
	30.0	0.311						
Equal interest in parameters and departures from all models	7.95	0.437						
	23.0	0.290	57.5	61.8	73.2	71.6	65.4	85.2
	30.0	0.273						

A disadvantage of 1 is that, with several models, the combined model may contain so many terms that the resulting optimum design, with many points of support, is far from optimum for any single departure.

***Example* 21.5** Three growth models

The methods of this section extend straightforwardly to non-linear models linearized, as in Chapter 18, by series expansion. To illustrate this together with multi-purpose designs for three models we consider the first-order growth model and the two alternatives to it which are given in Table 21.8. The inverse polynomial model, like the first-order growth model rises from 0 to 1 as t goes from 0 to ∞. The third possibility is a quadratic model without a constant. This might serve as a useful approximate model if the response were asymptotic to a value other than unity.

To detect departures from the two non-linear models, quadratic terms in $\partial\eta_1/\partial\theta$ and $\partial\eta_2/\partial\phi$ respectively can be added. For detecting departures from the quadratic model a term in t^3 is added. None of these generalizations yields designs efficient for determing whether the response is zero at $t=0$. This information could be provided, for example, by including in Table 21.8 a polynomial model with non-zero intercept.

A few equal-interest D-optimum designs are listed in Table 21.9. In calculating the designs the parameter θ for the first-order growth model was set to 0.1, so that all trials for the locally D-optimum design for estimating θ would be at $t=10$. The value of ϕ was chosen to make η_2 equal to η_1 at this point. For both non-linear models the design region can be unbounded. This is not the case for the quadratic model, and an upper value of 30 was taken for t. Three of the designs in Table 21.9 have three points of support: those for detecting cubic departures from the quadratic, those for equal interest in the specific departures from each model given in Table 21.8, and those for equal interest in the departures and the parameters. On the other hand, the combined model formed from the three individual models has four parameters. The equal-interest design for the complements of the three models in this combined model accordingly has four points of support. This design, generated from the extension of (21.19) to three models, is in general less efficient than the other equal-interest design for model checking and for parameter estimation. □

21.6 Further reading

Fuller details of the examples of equal-interest D-optimum designs are given by Atkinson (1975) and Atkinson and Cox (1974), which also includes a discussion of general ideas. Atkinson and Fedorov (1975a,b) describe several generalizations of D-optimality which are appropriate for non-sequential designs for discriminating between models. The Bayesian approach of

Spezzaferri (1988) reduces to D-optimality for some cases of vague prior information.

Sequential D-optimum designs for discrimination between models are described by Atkinson and Cox. With three or more models all methods so far considered involve some arbitrariness. One possibility is to modify the weights $a_i = 1/\tilde{s}_i$ of the generalization of the equal-interest design (21.19). If $R_i^{(n)}$ is the residual sum of squares for the ith model after n trials, for the $(n+1)$th trial take

$$a_i^{(n+1)} = \frac{1}{\tilde{s}_i} \exp\left(-\frac{1}{2} \frac{R_i^{(n)}}{R_{\min}^{(n)}} \right), \qquad (21.21)$$

where $R_{\min}^{(n)}$ is the minimum residual sum of squares amongst the v models. Even if Bayesian methods are used to replace (21.21) by the posterior probabilities of the models, the arbitrariness is not entirely resolved (Atkinson 1978; Smith and Spiegelhalter 1980; Spiegelhalter and Smith 1982).

Sequential designs for model-checking by augmentation and parameter estimation are conceptually closely related to those for discrimination between models. As it becomes clear whether the original or the augmented model holds, so the design weights should move from one aspect to another. Finally, we note that, although we have only considered generalizations of D-optimality in this chapter, designs combining several of the criteria of Chapter 10 are theoretically possible.

22
Further topics

22.1 Introduction

Although the topics covered in this chapter receive relatively brief treatment, this is not because they are of little practical importance. Indeed, the design of experiments for quality described in the next section has aroused appreciable interest in manufacturing industry, often under the name 'Taguchi methods'. One purpose of the experiments is to reduce the variability in performance of a product which is caused by variability in the environment in which the product is used. The requisite changes in the product are revealed by planned experiments which include environmental factors. A second application is to design products to be robust to fluctuations in the manufacturing process, a procedure sometimes called 'off-line quality control'. The distinction is with 'on-line control', where production is monitored to remove substandard product. Off-line control removes the wastage caused by rejection of substandard product and by the effort in detecting it. The experimental designs for both purposes are often simple when compared with those of earlier chapters. Some suitably simple first-order designs are listed in §22.3.

The designs of the following two sections are more complicated. In §22.4 optimum design is used for the sequential allocation of treatments to patients in a clinical trial. The method combines the iterative algorithm for the construction of optimum designs with randomization to provide allocations which have some balance but which also guard against biases in treatment allocation. The designs of the succeeding §22.5 have more in common with those of earlier chapters, but the methods are now extended to include generalized linear models. Designs are thus found for responses which might, for example, follow the gamma, Poisson, or binomial distributions. The examples given are for binary data.

In the final section brief mention is made of design problems arising in control theory and in agricultural experiments when account is taken of the potential correlation between yields on adjacent experimental units. The chapter ends with some aspects of the design of computer experiments.

22.2 Design for quality

A good photographic film will provide a sharp non-grainy image with true colours when used and developed under the conditions specified by the manufacturer. In this section a quality product is taken to be one which performs well under a variety of conditions. Thus, according to this definition, a quality film will provide a good image even if the exposure is incorrect and the developing solutions are at the wrong concentration and temperature. The experimental strategy for §3.3 for the optimization of yield would lead, in this example, to a film with good average properties. This section is concerned with the extension to experimental designs leading to quality products with high average performance which are in addition robust to environmental or usage factors.

It is convenient to divide the experimental factors into two groups. The design factors are the variables which have been the subject of most of the designs in earlier chapters. They have been varied, for example in D-optimum designs, to provide estimates of the parameters of the model and, in response surface designs of §3.3, to yield conditions of optimum yield. But now to these are to be added environmental factors which are also included in the experimental design.

A common form of design (e.g. Taguchi 1987) is the product of simple factorials, or their fractions, in the two sets of variables. Often the factorial for the design factors is called an inner array. This generates a series of trial products, each of which is then assessed for performance under the conditions of the design in the environmental variables, which form an outer array.

Although the idea of such designs is simple, interesting points arise both in design and analysis. These are illustrated with an example.

Example 22.1 Cake mix

This example, taken from Box *et al.* (1989), concerns the development of a new cake mix for the consumer market. The product needs to be robust to incorrect cooking conditions, represented by the environmental factors oven temperature x_4 and baking time x_5. The three design factors, which are under the control of the manufacturer, are the amounts of flour, sugar, and egg powder, denoted x_1, x_2, and x_3 respectively. The experimental design, given in Table 22.1, consists of a 2^3 factorial with centre point in the design factors crossed with a 2^2 factorial plus centre point in the environmental factors. The factor levels 0 in the table correspond to the intended composition of the mix and the suggested cooking conditions. The response is a taste index with large values desired. In passing we note that the variables in the three design factors would more properly be investigated using a mixture experiment.

Inspection of the results of Table 22.1 suggests that trials 7 and 9 produce mixtures which are least susceptible to variation in the environmental factors x_4 and x_5, but that trial 7 has the higher average and so is the best mix for the market.

This informal analysis of the results of this experiment suffices to extract the relevant information. However, in more complicated experiments of this type,

Table 22.1. Example 22.1: cake mix data. Design and average taste index

Design factors			Environmental factors					
Amounts of								
Flour	Sugar	Egg						
			x_4 Oven temperature	0	–	+	–	+
x_1	x_2	x_3	x_5 Baking time	0	–	–	+	+
0	0	0		6.7	3.4	5.4	4.1	3.8
–	–	–		3.1	1.1	5.7	6.4	1.3
+	–	–		3.2	3.8	4.9	4.3	2.1
–	+	–		5.3	3.7	5.1	6.7	2.9
+	+	–		4.1	4.5	6.4	5.8	5.2
–	–	+		5.9	4.2	6.8	6.5	3.5
+	–	+		6.9	5.0	6.0	5.9	5.7
–	+	+		3.0	3.1	6.3	6.4	3.0
+	+	+		4.5	3.9	5.5	5.0	5.4

a more sophisticated analysis may be required. As an example of the considerations involved, a more formal analysis is now given of the results of Table 22.1.

For the moment we maintain the distinction between design and environmental factors, treating the effect of each environmental factor as being to generate variance in the response. Table 22.2 gives, for each of the nine combinations of the design factors, the mean response $\bar{y}$ and the estimated variance s_y^2, together with $\log s_y^2$. These results confirm the informal impression of the robustness of mixes 7 and 9 to the environmental factors. In contrast, mix 6 has a high mean score, but also a large variance, so that it might produce unsatisfactory cakes if the cooking instructions were not followed sufficiently scrupulously.

The results of Table 22.2 come from a 2^3 factorial plus centre point. Table 22.3 gives the parameter estimates from the analysis of this design for each of the three responses $\bar{y}$, s_y^2 and $\log s_y^2$. Because there is only one degree of freedom for error, confounded with the lack-of-fit degree of freedom, we use half-normal plots of effects, described in §8.3, to indicate which effects are important.

The plot for the means (Fig. 22.1) shows that the important effects, in order, are -23, 3, and 1. This suggests that x_3 should be at the high level, x_2 at the low level, and x_1 at the high level in order to obtain the maximum average response, which correctly identifies mix 7. The half-normal plot for the estimated variances (Fig. 22.2) suggests that x_1 is the important factor for variance, and that it should be at its higher level. However, the plot seems not

Table 22.2. Example 22.1: cake mix. Summary statistics for the effects of environmental factors

x_1	x_2	x_3	$\bar{y}$	s_y^2	$\log s_y^2$
0	0	0	4.68	1.84	0.608
−	−	−	3.52	6.00	1.792
+	−	−	3.66	1.15	0.142
−	+	−	4.74	2.19	0.783
+	+	−	5.20	0.87	−0.134
−	−	+	5.38	2.12	0.750
+	−	+	5.90	0.46	−0.766
−	+	+	4.36	3.30	1.195
+	+	+	4.86	0.44	−0.814

Table 22.3. Example 22.1: cake mix. Estimated effects of design factors on the statistics of Table 22.2

Effect	$\bar{y}$	s_y^2	$\log s_y^2$
Mean	4.700	2.043	0.395
1	0.203	−1.334	−0.761
2	0.088	−0.366	−0.111
3	0.423	−0.486	−0.277
12	0.038	0.291	0.030
13	0.053	0.206	−0.120
23	−0.603	0.657	0.210
123	−0.043	−0.593	−0.153

to go through the origin. This is hardly surprising since the estimates of variance are distributed as scaled χ_4^2 random variables. A much closer approximation to normality of the effects is obtained from the analysis of $\log s_y^2$ (Fig. 22.3). This plot confirms that x_1 is the overridingly important factor in determining the variance. The remaining effects appear to fall on a straight line through the origin. This closer approach to normality comes from the symmetrizing effect on the positively skewed χ^2 distribution of taking logarithms (see, for example, McCullagh and Nelder 1989, p. 289).

These analyses have maintained the distinction between the design factors and the environmental factors. An alternative aproach, in line with the implicit analyses for the designs in the rest of this book, is to treat the design as a 2^5

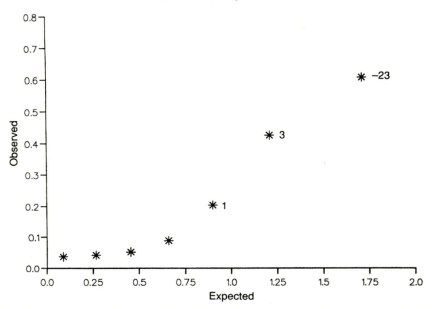

Fig. 22.1. Example 22.1: cake mix. Half-normal plot of effects of design variables on mean response averaged over environmental factors.

factorial with a variety of centre points. Figure 22.4 gives a half-normal plot of the 31 resulting effects, standardized for the slight differences in variance caused by the presence of the non-factorial points. This plot picks out the interaction of the environmental factors x_4 and x_5 as being of greatest importance, followed by the x_2x_3 interaction and two others. However, this combined analysis is of little use in suggesting a product which is robust to the environmental factors, whilst having a good taste. Although it might suggest different recommended values for the cooking conditions x_4 and x_5, the separation of the factors into two groups and the responses into mean and variance due to environmental factors are crucial to the understanding of this kind of experiment. □

The analysis of the cake mix experiment is straightforward because there is no conflict between the design conditions which increase average taste and those which decrease the variance due to the environmental factors. The high level of x_1 is needed to keep the variance down. To increase the mean, x_3 should be at the high level, x_2 at the low level, and x_1 again at the high level. Sometimes, however, there will be a conflict between design conditions which satisfy a target value for the mean response and those which are robust to environmental factors. In such cases much use has been made of so-called

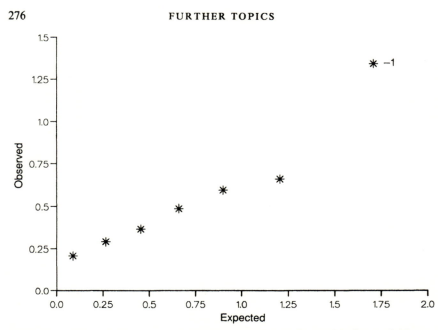

Fig. 22.2. Example 22.1: cake mix. Half-normal plot of effects of design variables on variance of the response over the enviromental factors.

signal-to-noise ratios of the form $\log(\bar{y}^2/s_y^2)$. If the effects of the desgin factors are as well behaved as they are for the cake mix example, maximization of the signal-to-noise ratio leads to simultaneous maximization of $\bar{y}$ and minimization of s_y^2. But, in general, maximization of the ratio will lead to maximization of the arbitrary linear combination $2 \log \bar{y} - \log s_y^2$. It will usually be better, when possible, to associate costs with both differences of $\bar{y}$ from the target value and with the variability s_y^2, and then to treat the index as a new response to be maximized. Another possibility is that of transforming the data. We saw in Chapter 8 that one of the indications of the desirability of a transformation of y was a relationship between mean and variance, which could be broken by a suitable transformation. Here too, although the circumstances are different, the relationship between mean and variance may sometimes be broken by transformation of the original response before calculation of means and variances. Box *et al.* (1989, p. 370) give a discussion, as do Logothetis and Wynn (1989, p. 253).

The cake mix example, and the introductory example on photographic film, are close to the standard technological applications of experimental design in the remainder of this book. But the methods of experimental design for quality are also of great importance in engineering design, where some new features arise.

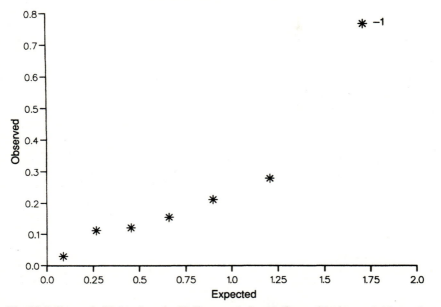

Fig. 22.3. Example 22.1: cake mix. Half-normal plot of effects of design variables on log variance of the response over the environmental factors.

Consider the design of an electrical circuit containing several components, an example given by Logothetis and Wynn (1989, p. 335). The design factors are the type, number, and arrangement of the various components. The environmental factors might include such external variables as temperature and humidity, but will also include the tolerances on the performances of the components. For example, a resistor with a nominal resistance of ρ ohms may have an actual resistance R which is normally distributed about ρ with a variance σ^2. For any particular design and specific values of random variables such as R, the performance of the system can be calculated from a mathematical model using well-established physical laws. Then there is no need of physical experimentation. The experiment, similar to that of Table 22.1, has as its inner array the results generated from the model at high and low values of the performance of each component, for example $\rho \pm 2\sigma$ or $\rho \pm 3\sigma$. These results are then analysed to yield understanding of the variability of the performance of the circuit as a function of variability in the components. This analysis may indicate a good design among those generated by the design factors. But it will also indicate which components are crucial to the performance of the circuit and perhaps suggest new target specifications for both ρ and the variance σ^2. If a variance is unnecessarily small, a cheaper component might be satisfactory. If, on the other hand, variability in the

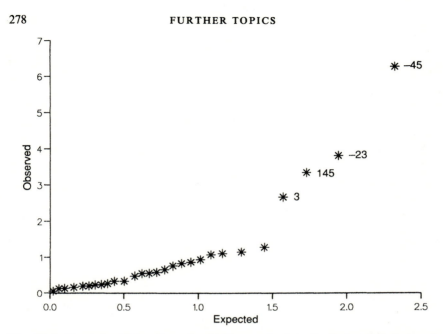

Fig. 22.4. Example 22.1: cake mix. Half-normal plot of effects of design and environmental factors from analysis as a full 2^5 factorial.

performance of a few components dominates the performance of the circuit, it will be worth tightening these specifications.

Of course, not all problems in engineering design can be solved by computer modelling. The paper feed of a photocopier or laser printer needs to be robust to variations in paper weight, quality, and size and to the effects of the storage conditions of the paper. Various designs of feed need to be compared using physical experiments. But whether the experiments take place in a computer or a laboratory, it is the patient and systematic accumulation of knowledge from experiments that leads to the continuing improvement in manufactured products that is an outstanding positive feature of the modern world. (Consider how much better your car is than the one your parents had at your age, or how your compact disc player compares with their gramophone). In particular, the application of the results of continuing purposeful experimentation is an important factor in the success of Japanese manufacturers. This approach to engineering design is given a systematic exposition by Logothetis and Wynn (1989). Three books of selected papers describing the methods outlined in this section and the accumulation of understanding and misunderstanding associated with the phrase 'Taguchi methods' are Barnard *et al.* (1989), Bendell *et al.* (1989), and Dehnad (1989). Finally, we note that, from the experimental design perspective of this book much remains to be done to improving the seemingly rather crude designs, such as that of Table 22.1, found by crossing factorials. The use of fractional factorials would obviously lead to a reduction in experimental effort. But if the purpose of the experiment is to explore regions of maximum average yield and low variance, second-order designs will be important in identifying better products and improved recommended conditions of use.

22.3 First-order designs

Designs for quality of the kind illustrated in §22.2 are often generated by crossing two first-order designs. The assumption for these designs to be appropriate is that interactions within the groups of factors are negligible. In this section we list some first-order designs which are sufficiently small to be useful.

The first appearance of first-order designs in this book was in §3.3 when they were introduced as screening designs. The example referred to there was the 2^{6-3} fractional factorial of Table 7.3, generated by assigning factors at two levels to the interaction columns of a 2^3 factorial. This design is extended in Table 22.4 to include an extra factor, giving a design for seven factors in eight trials, the maximum possible. Of course, when the design is used, the order of the trials should be randomized, as can be the allocation of factors to columns. If fewer than seven treatments are required, an arbitrary column of the design can be dropped, provided that all interactions can continue to be ignored. The extra degrees of freedom can then be used to estimate error. If, however, a particular interaction is of interest, the corresponding column of the design cannot have a factor associated with it.

Table 22.4. First-order design for up to seven factors in eight trials

Trial	Factor						
	1	2	3	4	5	6	7
1	−	−	−	+	+	+	−
2	+	−	−	−	−	+	+
3	−	+	−	−	+	−	+
4	+	+	−	+	−	−	−
5	−	−	+	+	−	−	+
6	+	−	+	−	+	−	−
7	−	+	+	−	−	+	−
8	+	+	+	+	+	+	+

The method of construction yielding Table 22.4 only works when N is a power of 2. Plackett and Burman (1946) provide orthogonal designs for factors at two levels for values of N which are multiples of 4 up to $N=100$, with the exception of the design for $N=92$, for which see Baumert *et al.* (1962).

The Plackett–Burman designs are found by the cyclical shifting of a generator which forms the first row of the design. Thus, for $N=12$, the generator is

$$+ \; + \; - \; + \; + \; + \; - \; - \; - \; + \; - \qquad (22.1)$$

which specifies the levels of up to 11 factors. The second row of the design is, as shown in Table 22.5, found by moving (22.1) one position to the right. Continuation of the process generates 11 rows. The 12th row consists of all factors at their lowest levels. Equivalent designs are found by reversing the $+$ and $-$ signs and by permuting rows and columns. The design should again be randomized before use.

Table 22.5. First-order design for up to 11 factors in 12 trials

Trial	Factor										
	1	2	3	4	5	6	7	8	9	10	11
1	+	+	−	+	+	+	−	−	−	+	−
2	−	+	+	−	+	+	+	−	−	−	+
3	+	−	+	+	−	+	+	+	−	−	−
4	−	+	−	+	+	−	+	+	+	−	−
5	−	−	+	−	+	+	−	+	+	+	−
6	−	−	−	+	−	+	+	−	+	+	+
7	+	−	−	−	+	−	+	+	−	+	+
8	+	+	−	−	−	+	−	+	+	−	+
9	+	+	+	−	−	−	+	−	+	+	−
10	−	+	+	+	−	−	−	+	−	+	+
11	+	−	+	+	+	−	−	−	+	−	+
12	−	−	−	−	−	−	−	−	−	−	−

The Plackett–Burman generators include values of N which are powers of 2, such as 16 and 32, and so provide an arguably easier alternative to generation than use of the 2^m factorial. The 16-trial design of Table 22.6 for 15 factors was, however, generated from the 2^4 factorial. Reversing $+$ and $-$ signs and permuting the rows and columns leads to the design given by Plackett and Burman with generator

$$+ \; + \; + \; + \; - \; + \; - \; + \; + \; - \; - \; + \; - \; - \; - . \qquad (22.2)$$

The final design given in full here is the design for $N = 20$ of Table 22.7. This, at least in our experience, seems to be about the limit of practical applications. On the off-chance that it might be needed, the generator for $N = 24$ is

$$+ + + + + - + - + + - - + + - - - + - + - - - - . \quad (22.3)$$

Since the design matrices for all these designs are orthogonal, the effect of each factor is estimated as if it were the only one in the experiment. Therefore it follows that the designs are D-optimum.

Table 22.6. First-order design for up to 15 factors in 16 trials

Trial	Factor														
	1	2	3	4	5	6	7	8	9	10	11	12	13	14	15
1	−	−	−	−	+	+	+	+	+	+	−	−	−	−	+
2	+	−	−	−	−	−	−	+	+	+	+	+	+	−	−
3	−	+	−	−	−	+	+	−	−	+	+	+	−	+	−
4	+	+	−	−	+	−	−	−	−	+	−	−	+	+	+
5	−	−	+	−	+	−	+	−	+	−	+	−	+	+	−
6	+	−	+	−	−	+	−	−	+	−	−	+	−	+	+
7	−	+	+	−	−	−	+	+	−	−	−	+	+	−	+
8	+	+	+	−	+	+	−	+	−	−	+	−	−	−	−
9	−	−	−	+	+	+	−	+	−	−	−	+	+	+	−
10	+	−	−	+	−	−	+	+	−	−	+	−	−	+	+
11	−	+	−	+	−	+	−	−	+	−	+	−	+	−	+
12	+	+	−	+	+	−	+	−	+	−	−	+	−	−	−
13	−	−	+	+	+	−	−	−	−	+	+	+	−	−	+
14	+	−	+	+	−	+	+	−	−	+	−	−	+	−	−
15	−	+	+	+	−	−	−	+	+	+	−	−	−	+	−
16	+	+	+	+	+	+	+	+	+	+	+	+	+	+	+

The Plackett–Burman designs are limited by the requirement that N be a multiple of 4. Although this is not usually an important practical constraint, the resulting mathematical problem of finding designs for other N has attracted appreciable attention. One possibility is to combine computer searches for optimum designs of the kind described in earlier chapters with the use of orthogonal arrays. Mitchell (1974*b*) illustrates the use of DETMAX in finding first-order designs when N is not a multiple of 4 and discusses separately the properties of designs for the three cases $N = 4t + 1$, $N = 4t + 2$, and $N = 4t + 3$, where t is an integer. The most difficult case, when $N = 4t + 3$, is investigated by Galil and Kiefer (1982). For more recent references see Atkinson (1988).

Table 22.7. First-order design for up to 19 factors in 20 trials

Trial	Factor																		
	1	2	3	4	5	6	7	8	9	10	11	12	13	14	15	16	17	18	19
1	+	+	−	−	+	+	+	+	−	+	−	+	−	−	−	−	+	+	−
2	−	+	+	−	−	+	+	+	+	−	+	−	+	−	−	−	−	+	+
3	+	−	+	+	−	−	+	+	+	+	−	+	−	+	−	−	−	−	+
4	+	+	−	+	+	−	−	+	+	+	+	−	+	−	+	−	−	−	−
5	−	+	+	−	+	+	−	−	+	+	+	+	−	+	−	+	−	−	−
6	−	−	+	+	−	+	+	−	−	+	+	+	+	−	+	−	+	−	−
7	−	−	−	+	+	−	+	+	−	−	+	+	+	+	−	+	−	+	−
8	−	−	−	−	+	+	−	+	+	−	−	+	+	+	+	−	+	−	+
9	+	−	−	−	−	+	+	−	+	+	−	−	+	+	+	+	−	+	−
10	−	+	−	−	−	−	+	+	−	+	+	−	−	+	+	+	+	−	+
11	+	−	+	−	−	−	−	+	+	−	+	+	−	−	+	+	+	+	−
12	−	+	−	+	−	−	−	−	+	+	−	+	+	−	−	+	+	+	+
13	+	−	+	−	+	−	−	−	−	+	+	−	+	+	−	−	+	+	+
14	+	+	−	+	−	+	−	−	−	−	+	+	−	+	+	−	−	+	+
15	+	+	+	−	+	−	+	−	−	−	−	+	+	−	+	+	−	−	+
16	+	+	+	+	−	+	−	+	−	−	−	−	+	+	−	+	+	−	−
17	−	+	+	+	+	−	+	−	+	−	−	−	−	+	+	−	+	+	−
18	−	−	+	+	+	+	−	+	−	+	−	−	−	−	+	+	−	+	+
19	+	−	−	+	+	+	+	−	+	−	+	−	−	−	−	+	+	−	+
20	−	−	−	−	−	−	−	−	−	−	−	−	−	−	−	−	−	−	−

Designs for factors at more than two levels are not so easily found. Plackett and Burman give generators for some designs with all factors at three levels. Tables of fractions of three-level factorials are given by Connor and Zelen (1959). Mitchell and Bayne (1978) again combine the properties of orthogonal arrays with the use of DETMAX to search for fractional factorials. Such designs are closely related to the D-optimum exact designs for second-order models of Chapter 15 found by searching over the points of the 3^m factorial.

22.4 Biased coin designs for clinical trials

Patients in a clinical trial arrive sequentially and are assigned to one of t treatments. Because it is not known when the trial will terminate, this assignment should, as the trial evolves, maintain a balance between the numbers receiving each treatment. But, if balance is strictly maintained, it is often possible to guess, with a high probability of success, which treatment the next patient will receive. Therefore the allocation also needs to be sufficiently random to avoid any suspicion of conscious or unconscious cheating. To balance these requirements Efron (1971) introduced biased coin designs for the comparison of two treatments. In such designs the allocation of the treatment is determined probabilistically, but with a fixed bias towards the under-represented treatment.

One disadvantage of Efron's scheme is that it does not include balance over covariates or prognostic factors which may affect the response of the patient to the treatment. In this section the results on D_A-optimality of §10.2 are used, following Atkinson (1982b), to generate biased coin designs for any number of treatments which allow the possibility of balance over prognostic factors. The resulting sequential allocation depends on the data through the values of earlier treatment allocations and prognostic factors. However, it does not depend on the responses from earlier patients.

Implicit behind the procedure is the ususal linear model $E(Y) = F\beta$ for the response of the patient. The treatment effects are represented by t indicator variables, with an unspecified linear model for the prognostic factors. For this experiment the design region $\mathscr{X}$ has a simple structure, consisting of t points, the ith of which represents allocation of treatment i to the next patient. Interest in the analysis of clinical trials often centres on contrasts between treatment effects. Suppose that the contrasts are s linear combinations which are elements of the vector $A^T\beta$, where A is a $p \times s$ matrix of rank $s < p$. As in §10.2, the covariance matrix of the least squares estimate $A^T\hat{\beta}$ is proportional to $A^T M^{-1}(\xi)A$ and the D_A-optimum design minimizing $\log |A^T M^{-1}(\xi)A|$ has derivative function $d_A(x, \xi)$ given by (10.2). Let the value of $d_A(x, \xi)$ for allocation of treatment i be $d_A(i, \xi)$. The standard iterative construction of optimum continuous designs allocates the $(n+1)$th patient to the treatment for which $d_A(i, \xi_n)$ is a maximum. At the optimum design, which allocates a fraction f_i^* of the patients to treatment i, all $d_A(i, \xi^*)$ are equal. A large value of $d_A(i, \xi_n)$ indicates a treatment which is under-represented.

It may be that in trials with several centres and clinicians the order of arrival of patients provides sufficient randomization to guard against selection bias. If this is the case, the iterative construction of the optimum design can be used directly. But, if randomization is required, we propose a biased coin design in which treatment i is chosen with probability

$$\pi_i = \frac{f_i^* d_A(i, \xi_n)}{\Sigma_{i=1}^t f_i^* d_A(i, \xi_n)}. \tag{22.4}$$

This has the property that for the optimum design when $\xi_n = \xi^*$, each treatment is selected with probability f_i^*. If the design departs from the optimum, the allocation attempts to reduce the departure.

The values of the optimum fractions f_i^* depend on the matrix A. If interest is in the $(t-1)$-dimensional space of all contrasts orthogonal to the mean, the D_A-optimum design in the absence of prognostic factors allocates an equal number of patients to each treatment. If prognostic factors are present, the same design is optimum under regularity conditions which would asymptotically ensure the identical distribution of prognostic factors to patients receiving each treatment. In either case $f_i^* = 1/t$ for all treatments. The biased coin rule (22.4) then reduces to choosing treatment i with probability

$$\pi_i = \frac{d_A(i, \xi_n)}{\Sigma_{i=1}^t d_A(i, \xi_n)}. \tag{22.5}$$

When $\xi_n = \xi^*$ the design is balanced and each treatment is selected with equal probability.

An arbitrary feature of these definitions of π_i is that $d_A(i, \xi_n)$ could, with equal justification, be replaced by any monotone function of itself. In theory the monotone function should be chosen to balance the loss due to selection bias against the loss due to imbalance in treatment allocation. This possibility will not be explored here. Instead, we describe three cases of the general method in order of increasing complexity.

***Example* 22.2** Two treatments, no prognostic factors

In the absence of prognostic factors the model for two treatments reduces to $E(Y) = \beta_i$ $(i = 1, 2)$. If, as is usually the case, interest is in the difference between treatment effects $\beta_1 - \beta_2$, $A^T = (1 \ -1)$ and $s = 1$. If n_1 of the n patients have received treatment 1 and n_2 have received treatment 2, than the information matrix $M(\xi_n) = \text{diag}\{n_i/n\}$. Substitution in (10.2) shows that

$$d_A(1, \xi_n) = n_2/n_1 \qquad d_A(2, \xi_n) = n_1/n_2.$$

From the biased coin rule (22.5), the probability of selecting treatment 1 for the $(n+1)$th patient is

$$\pi_1 = \frac{n_2^2}{n_1^2 + n_2^2}, \tag{22.6}$$

with the complementary expression for π_2.

This D_A-optimum design has been derived on the assumption that only the difference in treatments is of interest. If the mean treatment effect is also relevant, the D-optimum design should be used, for which the derivative $d(x, \xi_n) = n/n_i$ $(i = 1, 2)$. If these variances are used as the basis of a biased coin rule of the kind considered here,

$$\pi_1' = n_2/n \qquad \pi_2' = n_1/n. \tag{22.7}$$

For both D- and D_A-optimum designs the optimum allocation is to have the same number of patients receiving each treatment, i.e. $n_1 = n_2 = 1/2$. The equivalence theorems confirm this obvious result since $d_A(i, \xi^*) = 1 = s$ and $d(i, \xi^*) = 2 = p$, the number of parameters in the linear model.

Some numerical values of the allocation probabilities π_1 and π_1' are given in Table 22.8 for various values of n and of the 'imbalance' defined as $n/2 - n_1$.

Both rules derived from optimum design theory provide a 'restorative force' which increases as the design moves away from balance. This contrasts with the original suggestion of Efron (1971) in which, if $n_2 > n_1$, treatment 1 is chosen with probability greater than 1/2, with 2/3 Efron's favourite value. The results of Table 22.8 show that the design based on D_A-optimality responds much more to imbalance than does the D-optimum-based design. For $n = 10$ an imbalance of 1 gives a value of $\pi_1 > 2/3$ but, as n increases, larger imbalances are needed to achieve values comparable with Efron's rule, a result which follows from expressing π_1 as a function of the ratio n_2/n. ☐

Example 22.3 Three or more treatments, no prognostic factors

With two treatments the form of the matrix A^T is obvious. With three or more treatments, when all contrasts between the treatments are of equal interest, the form of A^T is arbitrary provided that the $s = t - 1$-dimensional space of the contrasts is orthogonal to the overall mean. In the absence of prognostic factors, A^T is irrelevant once the values of $d_A(i, \xi_n)$ have been obtained. However, an explicit form of A^T is needed for the calculation of designs with prognostic factors and one possible form is now derived.

Let

$$A^T = \begin{bmatrix} 1 & -1 & 0 & \cdots & 0 \\ 1 & 0 & -1 & \cdots & 0 \\ \vdots & \vdots & \vdots & & \vdots \\ 1 & 0 & 0 & & -1 \end{bmatrix}$$

so that the set of $t - 1$ contrasts $\beta_1 - \beta_i$ is considered for $i = 2, \ldots, t$. If $f_i = n_i/n$, the information matrix is $M(\xi_n) = \text{diag}\{f_i\}$. If, further, $r_i = 1/f_i = n/n_i$,

$$A^T M^{-1}(\xi) A = r_1 J + \text{diag}\{r_2, \ldots, r_t\}$$

where J is the $(t-1) \times (t-1)$ matrix with all elements 1. The matrix identity

$$(C + uv^T)^{-1} = C^{-1} - \frac{C^{-1}uv^T C^{-1}}{1 + v^T C^{-1} u},$$

where u and v are vectors, leads to the result

$$\{A^T M^{-1}(\xi) A\}^{-1} = \begin{bmatrix} f_2(1 - f_2) & -f_2 f_3 & \cdots & -f_2 f_t \\ -f_2 f_3 & f_3(1 - f_3) & \cdots & -f_3 f_t \\ \vdots & & & \vdots \\ -f_2 f_t & -f_3 f_t & \cdots & f_t(1 - f_t) \end{bmatrix}, \quad (22.8)$$

Table 22.8. Biased coin design for clinical trials. Probability of assigning treatment 1 under three different schemes, when there are two treatments and no prognostic factors

Imbalance $N/2 - n_1$	D_A-optimum π_1 (22.6) Number of trials N				D-optimum π_1 (22.7) Number of trials N				Efron N unspecified
	10	20	50	100	10	20	50	100	
1	0.692	0.599	0.540	0.520	0.6	0.55	0.52	0.51	0.667
2	0.845	0.692	0.575	0.540	0.7	0.60	0.54	0.52	0.667
3	0.941	0.775	0.618	0.560	0.8	0.65	0.56	0.53	0.667
4	0.988	0.845	0.656	0.579	0.9	0.70	0.58	0.54	0.667
5	1	0.900	0.692	0.599	1	0.75	0.60	0.55	0.667
6		0.941	0.727	0.618		0.80	0.62	0.56	0.667

from which it follows that $d_A(i, \xi_n) = (n - n_i)/n_i = r_i - 1$, with r_i the reciprocal of the proportion of patients receiving treatment i.

The D_A-optimum design again consists of equal replication and $d_A(i, \xi^*) = t - 1$. The biased coin design therefore has probabilities given by

$$\pi_i = \frac{r_i - 1}{\sum_{i=1}^{t} r_i - t},$$

which is the appropriate generalization of (22.6) to more than two treatments.

□

Example 22.4 Several treatments with prognostic factors

To include prognostic factors, write the model in the partitioned form

$$E(Y) = F\beta = E\alpha + Z\gamma, \qquad (22.9)$$

where E is the matrix of indicator variables for the treatments and Z is the matrix of the $p - t$ prognostic factors. Since the nuisance parameters γ are not of interest, we require the D_A-optimum design for the linear combination $L^T\beta$ where

$$L^T = (A^T\ 0),$$

with A as before identifying the $t - 1$ contrasts of interest. Experiments are thus being designed to estimate $A^T\alpha$ as precisely as possible, which can be thought of as a generalization of D_s-optimality.

Calculation of the design requires partition of the dispersion matrix according to (22.9) which yields

$$M^{-1}(\xi) = \begin{bmatrix} M^{11} & M^{12} \\ M^{21} & M^{22} \end{bmatrix}.$$

It is now no longer possible to obtain analytical expressions for the π_i. However, the calculation of $d_A(i, \xi_n)$ can be simplified. For the contrasts given by A^T above (22.8) the general expression (10.2) becomes

$$d_A(f, \xi_n) = e^T M^{11} B M^{11} e + 2e^T M^{11} B M^{12} z + z^T M^{21} B M^{12} z, \quad (22.10)$$

where $B = A(A^T M^{11} A)^{-1} A^T$ and e^T and z^T are rows of the matrices of variables E and Z.

Calculation of the quadratic form (22.10) is required for the sequential biased coin allocation of a treatment to each successive patient. But the calculation is straightforward, as is searching over the design region $\mathscr{X}$ to find the t values of $d_A(i, \xi_n)$. The three kernels of the quadratic form (22.10) depend upon past history and so can be calculated before the arrival of the next patient. The vector z is the set of prognostic factors for the new patient. Once these are known, (22.10) reduces to a quadratic form in e, together with a linear term and a constant. But e is the indicator variable for the treatment. Thus searching over $\mathscr{X}$ is achieved by picking successive diagonal elements of $M^{11}BM^{11}$ and successive terms of the vector $M^{11}BM^{12}z$. Once the t values $d_A(i, \xi_n)$ have been calculated, it only remains to find the probabilities π_i and to perform the biased coin allocation. $\qquad\qquad\qquad\qquad\qquad\qquad\qquad\qquad\qquad\quad$ □

The extension of the iterative construction of optimum designs used here in combination with randomness to introduce unpredictability into treatment allocation is not the only way of extending Efron's original biased coin designs. Wei (1978a,b) used urn designs for the extension to more than two treatments. Some properties of both Wei's designs and those of this section have been investigated. Smith (1984a) studies the distribution of the numbers of patients receiving each of several treatments when prognostic factors are present. Smith (1984b) applies the results to a variety of biases that could arise: selection bias, bias die to outliers, and accidental biases due to smooth trends and correlated errors. The problems of inference following the use of such designs are considered by Cox (1982), Smith (1984b), and Wei et $al.$ (1986), and by Begg (1990) and his discussants. Heckman (1985) gives a further local limit theorem for designs with two treatments.

22.5 Generalized linear models

The methods of experimental design described so far are appropriate if the response, perhaps after transformation, has a variance which is constant over the experimental region. This is a characteristic of the customary error assumptions made in applications of the normal theory linear model. However, it is not a characteristic of such highly non-normal error distributions as the binomial or Poisson, where there is a strong relationship between mean and variance. In this section the methods of optimum experimental design are extended to include both normal observations with non-constant variance and the more general case of non-normal responses.

The assumptions of normality and constancy of variance enter the optimum design through the information matrix $F^{\mathrm{T}}F$. Other forms of matrix are appropriate for other distributions. Once the appropriate matrix has been defined, the principles and practice of optimum experimental deisgn are similar to those described in earlier chapters. We shall mostly be concerned with the family of generalized linear models described by McCullagh and Nelder (1989), in which the asymptotic covariance matrix of the parameters of the linear model is of the form $F^{\mathrm{T}}WF$, where the $N \times N$ diagonal matrix of weights W depends upon the parameters of the linear model, the error distribution, and the link between them. This dependence on parameters means that, in general, designs for generalized linear models require the specification of prior information: point prior information will lead to locally optimum designs, analogous to the designs for non-linear models of Chapter 18. Full specification of a prior distribution leads to Bayesian optimum designs similar to those of Chapter 19. Because the structure of $F^T W F$ is that which is obtained for weighted least squares, we start with a short discussion of design for non-constant-variance normal theory linear models.

For the linear regression model $E(Y) = F\beta$ let the variance of an observation at x_i be $\sigma^2 / w(x_i)$ ($i = 1, \ldots, N$), where the $w(x_i)$ are a set of known weights. It is still assumed that the observations are independent. One special case is when the observation at x_i is $\bar{y}_i$, the mean of n_i observations, when $\text{var}(\bar{y}_i) = \sigma^2 / n_i$ so that $w(x_i) = n_i$. Efficient least squares analysis of such data makes use of the weights. If $W = \text{diag}\{w(x_i)\}$, the weighted least squares estimate of the parameter β is

$$\hat{\beta} = (F^T W F)^{-1} F^T W y$$

where

$$\text{var}(\hat{\beta}) = \sigma^2 (F^T W F)^{-1}. \tag{22.11}$$

The design criteria of Chapter 10 can be applied to the information matrix $F^T W F$ to yield optimum designs. In particular, the D-optimum exact design maximizes $|F^T W F|$. For continuous designs the information matrix is

$$M(w, \xi) = \int w(x) f(x) f(x)^T \xi(dx) \tag{22.12}$$

with the D-optimum design maximizing $|M(w, \xi)|$. The equivalence theorem for D-optimality follows by letting

$$\bar{d}(w, \xi) = \max_x w(x) f^T(x) M^{-1}(\xi) f(x), \tag{22.13}$$

with $\bar{d}(w, \xi^*) = p$.

Example 22.5 Weighted regression through the origin

The model is $E(Y) = x\beta$, a straight line through origin, with $x \geq 0$. If the variance of the errors is constant, the D-optimum design puts all trials at the maximum value of x. Suppose, however, that $\text{var}(Y_i) = \sigma^2 e^x$, so that the variance increases exponentially fast with x. In the notation of this section the weight $w(x) = e^{-x}$.

The D-optimum design for this one-parameter model concentrates all mass at one point. Since $f(x) = x$, the information matrix is

$$M(w, \xi) = x^2 e^{-x}, \tag{22.14}$$

which is a maximum when $x = 2$. Then

$$d(x, w, \xi^*) = e^{-x} x M^{-1}(w, \xi^*) x = x^2 \frac{e^{-x}}{4e^{-2}},$$

taking the value 1 when $x = 2$, which is the maximum over $\mathscr{X}$. Thus the D-optimum design has been found. The rapid increase of the variance as x increases leads to a design which does not span the experimental region, provided that the maximum value of x is greater than 2.   □

This example illustrates the effect of the weight $w(x)$. Otherwise the method of finding the design is the same as that for regression with constant variance. Since the fitting method for generalized linear models is weighted least squares, with the asymptotic covariance matrix of the parameter estimates given by (22.11), the methods of design for this family are again similar. The standard reference for generalized linear models is McCullagh and Nelder (1989), with Dobson (1990) providing a brief introduction. We start with a brief survey of basic ideas. A fuller treatment of the relevant material is given in Chapter 2 of McCullagh and Nelder.

The generalized linear model is an extension of the normal theory linear model to other distributions such as the gamma, Poisson, and binomial. The model has three components, which will be illustrated using the binomial distribution.

1. *The random component.* The N random variables Y are independently distributed with mean μ and

$$\text{var}(Y) = a(\phi)V(\mu). \tag{22.15}$$

In (22.15) $a(\phi)$ is a scale factor which does not depend upon μ. For the normal distribution $a(\phi) = \sigma^2$ and $V(\mu) = 1$; the mean and the variance are not related. However, they are related for the binomial distribution. Let R_i be the number of successes in n_i trials and set $Y_i = R_i/n_i$. Then, if $E(Y_i) = \mu_i$, $\text{var}(Y_i) = \mu_i(1 - \mu_i)/n_i$ so that $V(\mu) = \mu(1 - \mu)$ and $a(\phi) = 1/n_i$.

2. *The linear predictor.* The linear model provides the linear predictor

$$\eta = F\beta, \tag{22.16}$$

where F is again an $N \times p$ matrix of known functions of the m explanatory variables. The extension to non-linear models is to write $\eta = F(x, \theta)$, for which the methods are analogous to those of Chapter 18. This extension to generalized non-linear models is discussed by McCullagh and Nelder (1989, §11.4).

3. *The link function.* The final component of the generalized linear model is the link between the mean vector μ and the linear predictor, which is written

$$g(\mu) = \eta. \tag{22.17}$$

The correct link function for a particular set of data is a matter of empirical investigation. However, useful links are usually found to be those which map the unrestricted range of $F\beta$ to the range of μ. For multiple regression, positive and negative values of Y are usually possible, so that the identity link $g(\mu) = \mu$ is appropriate, leading to the standard regression model $E(Y) = \mu = F\beta$. However, for most generalized models, the value of μ is constrained. For the Poisson model it is physically necessary that $\mu \geqslant 0$. Then the logarithmic link $\log \mu = \eta$, or equivalently $\mu = e^{\eta}$, guarantees non-negativity of μ, whatever the values of F and β.

For binomial data, since Y_i is defined to be R_i/n_i, we require $0 \leqslant \mu_i \leqslant 1$. Three link functions that have been found useful are the following.

1. Logistic:

$$\eta = \log\left(\frac{\mu}{1-\mu}\right). \tag{22.18}$$

2. Probit:

$$\eta = \Phi^{-1}(\mu) \tag{22.19}$$

where Φ is the normal cumulative distribution function.

3. Complementary log–log:

$$\eta = \log\{-\log(1-\mu)\}. \tag{22.20}$$

The probit and logistic links have very similar properties (McCullagh and Nelder 1989, p. 109). The probit has the longer history (Finney 1971), dating back to the time when the normal distribution dominated the analysis of data. The interpretation of (22.18) as directly modelling logarithmic odds is an advantage and we shall not consider design for the probit link. For the analysis of binary data see Cox and Snell (1989).

The D-optimum design depends on all three components of the model. The distribution determines the variance function $V(\mu)$, which is one component of the weights W in (22.12). For generalized linear models the weights are given by

$$W = V^{-1}(\mu)\left(\frac{d\mu}{d\eta}\right)^2. \tag{22.21}$$

To illustrate optimum design for generalized linear models we evaluate W

for two different links for binary data and find the D-optimum design measures maximizing $|M(w, \xi)|$. Exact designs for binomial random variables with a total of N trials can be found by integer approximation to ξN. We do not try to find optimum exact N trial designs of the kind found in Chapter 15 for normal regression.

Example 22.6 Binary data with the logistic link

For binary data $V(\mu) = \mu(1 - \mu)$. To calculate W (22.21) requires the derivative $d\mu/d\eta$ or, equivalently from differentiation of the logistic link (22.18),

$$\frac{d\eta}{d\mu} = \frac{1}{\mu(1 - \mu)}, \tag{22.22}$$

leading to the relatively simple form

$$W = \mu(1 - \mu). \tag{22.23}$$

As an example, consider the linear model in one explanatory variable for which

$$\log\left(\frac{\mu}{1 - \mu}\right) = \beta_0 + \beta_1 x. \tag{22.24}$$

Since this model is linear in β, the information matrix (22.12), and so the D-optimum design, will depend on β only through W and so only through the value of μ. The values of the parameters for calculation of the design can then be chosen for convenience. We take $\beta_0 = 0$ and $\beta_1 = 1$. The D-optimum design is concentrated at two points, an assertion which can be checked by use of the equivalence theorem, and will therefore put equal weight at these two points. Because of the symmetry of (22.24) in x for our chosen parameter values, the two optimum values of x will be symmetrical about zero.

Let the weights when $\beta_0 = 0$ and $\beta_1 = 1$ be called $w(x, 0)$. From (22.24), $\mu = e^x/(1 + e^x)$ so that

$$w(x, 0) = \mu(1 - \mu) = \frac{e^x}{(1 + e^x)^2}$$

and $w(-x, 0) = w(x, 0)$. If the design points are a and $-a$, the information matrix is

$$M(w, \xi) = \frac{w(a, 0)}{2}\left\{\begin{pmatrix} 1 & -a \\ -a & a^2 \end{pmatrix} + \begin{pmatrix} 1 & a \\ a & a^2 \end{pmatrix}\right\} \tag{22.25}$$

so that the D-optimum design is found by maximizing $a^2 w^2(a, 0)$ or equivalently $aw(a, 0) = ae^a/(1 + e^a)^2$. The locally D-optimum design is then

$$\xi^* = \left\{ \begin{matrix} -1.5434 & 1.5434 \\ 0.5 & 0.5 \end{matrix} \right\}. \tag{22.26}$$

For this optimum design $\mu(a) = 0.8240$ and so $\mu(-a) = 1 - \mu(a) = 0.1760$. As with Example 22.5, it has not been necessary to limit $\mathscr{X}$: the effect of the increase of variance as the response tends to 1 or -1 with extreme x values is to ensure that the design avoids such values.

This result agrees with that of Chaloner (1988) who finds the locally D-optimum design for $\beta_0 = 0$ and $\beta_1 = 7$. For these parameter values the two design points are ± 0.2205. Since $7 \times 0.2205 = 1.5435$ it follows from (22.24) that the two designs give experiments at the same value of μ. $\qquad\square$

In addition to locally optimum designs, Chaloner and Larntz (1989) explore the properties of optimum Bayesian design for the logistic model, including D-optimum designs similar to those of Chapter 19 for non-linear regression models. In particular, they display a number of designs which demonstrate the increase in the number of design points with increasing prior uncertainty about the parameter values.

Example 22.7 Binary data with the complementary log–log link

Now suppose that the linear model for binary data is the same as before, but that the link is the complementary log–log (22.20), rather than the logistic of Example 22.6. Differentiation of (22.20) yields

$$\frac{d\eta}{d\mu} = \frac{1}{(1 - \mu)\log(1 - \mu)},$$

so that the weights are given by

$$W = \frac{1 - \mu}{\mu} \log^2(1 - \mu), \tag{22.27}$$

a more complicated form than the logistic weights (22.23). To find the locally optimum design we again take $\beta_0 = 0$ and $\beta = 1$. However, (22.27), unlike (22.20), is not symmetrical about $\mu = 1/2$. The locally D-optimum design found by numerical maximization of $M(w, \xi)$ is

$$\xi^* = \left\{ \begin{matrix} -1.338 & 0.9796 \\ 0.5 & 0.5 \end{matrix} \right\},$$

at which the values of μ are 0.2308 and 0.9303. Both values of μ are higher than those for the optimum design for the logistic model. In particular, the trials at $x = 0.9796$ yield a probability of success appreciably closer to unity than the 0.8240 of (22.6). $\qquad\square$

These two examples illustrate many of the similarities between the construction of D-optimum designs for generalized linear models and for regression models. As well as D-optimum designs for the binary logistic model, Chaloner and Larntz (1989) calculate examples of analogues of the c-optimum designs of §19.4. These are used to provide designs for the estimation of doses for specific values of μ; often interest is either in the LD_{50} or LD_{95}, i.e. the value of x for which $\mu = 0.5$ or $\mu = 0.95$. The extension of these procedures to any generalized linear model is outlined by Chaloner (1988).

The designs derived in this section assume that the form of the model is known. However, there is also the problem of discrimination between models. Chambers and Cox (1967) find designs for discrimination between the logistic and probit links with one explanatory variable. For the logistic null hypothesis the optimum measure puts about a third of the trials close to the D-optimum design (22.26). The remaining two-thirds go where $\mu = 0.9964$. To obtain any information at such an extreme dose clearly takes many trials. One calculation shows that about 4500 trials are needed to obtain a power of 95 per cent for a test of one model against the other. Since Chambers and Cox do not use optimum design theory, they are not able to use an equivalence theorem to demonstrate the optimality of their design. However, the more general approach to discrimination between models of Chapter 20, with its attendant equivalence theorem, can be adapted to generalized linear models. In the designs found there the non-centrality parameter to be maximized by design was the expected value of the residual sum of squares for the false model in the absence of error. The analogue for generalized linear models is to maximize the expected deviance. The details are given by Ponce de Leon (1992).

22.6 In brief

Despite the many applications of the theory of optimum experimental design that have been described in earlier chapters, there remain several important areas which have not been mentioned. Two can be thought of as arising from the extension of the information matrix $F^T W F$ to include non-diagonal W. One is design when the observations are a time series, so that successive readings are correlated. Titterington (1980) gives a review of work in control theory where one problem is to design experiments to estimate the parameters of a control system. Zarrop (1979) provides a book-length treatment of design for dynamic systems. References to more recent work are given by Rafajlowicz (1989). A second example of a non-diagonal covariance matrix arises through the use of neighbour methods for the analysis of field trials: the purpose is to allow for the covariance of yields on adjacent plots through modelling rather than by dividing the experiment into blocks. If the factors are qualitative, the resulting block designs, such as Latin squares, have constraints on the ocurrences of adjacent pairs of treatments. Amongst papers on design are

Bailey (1984), Street and Street (1985), Williams (1985), and Martin (1986), which includes an appreciable list of references. Kiefer and Wynn (1984) includes a brief discussion of design for the sequential clinical trials of §22.4 when there is correlation between successive responses.

A positive feature of the optimum design of experiments which has been stressed in this book is that the theory leads to algorithms which can readily be implemented on a computer. One implementation is given in Appendix A. It therefore seems appropriate to conclude with some remarks on the design of computer simulation experiments.

In the most common form of simulation a stochastic model is used to generate repeated realizations of a sample, for each of which a quantity of interest is calculated. For example, the aim may be to find the null distribution of a test statistic. Often the constraint on resources is not important: a more pressing problem frequently seems to be to find a cogent presentation of the results. For example, normal plots provide a clear summary of power simulations (Atkinson 1985, p. 168). However, if computing time is significant, several variance reduction techniques are available, such as the use of antithetic variables and importance sampling. Ripley (1987, Chapter 5) discusses experimental design and variance reduction. He makes the important point that there needs to be a balance between saving of computer time and the cost to the simulator of devising and programming a more efficient method.

A form of simulation which avoids parametric modelling is the bootstrap in which the original data are resampled. In its simplest form let the data be the random sample $Y_1, \ldots, Y_n$. Bootstrap samples are generated by sampling with replacement from $Y_1, \ldots, Y_n$ to give n^n possible replacement samples. Some of these will be far from the original sample and may increase the variance of the estimate of interest. One way of avoiding an excess of extreme samples is balanced resampling. In first-order balance each Y_i occurs the same number of times, say m, in the total of mn resampled observations. Davison et al. (1986) achieve this requirement by concatenating m copies of the original sample and then permuting the mn observations. Successive resamples are obtained by reading blocks of n of the permuted observations. This scheme leads to first-order balance, appropriate for the calculation of average values. Designs with second-order balance, to be used for the calculation of variances, are given by Graham et al. (1990). Another approach to obtaining approximate balance is that of Wynn and Ogbonmwan (1986) who view all possible resamples as points in an n-dimensional cube of side n. The coordinates of the point are the indices of the data points in the resample. The design problem is then that of choosing m points which uniformly fill the cube.

In both of these applications of experimental design to simulation the aim is to investigate and evaluate variability. However, many simulations are deterministic. Sacks et al. (1989) describe several computer models arising, for example, in studies of combustion. The controllable inputs are the parameters of the system, such as chemical rate constants and mixing rates. The deterministic response is hard to calculate, as it involves the numerical solution of sets of simultaneous partial differential equations. Because the response is deterministic, replication of a design point yields an identical result. Yet the choice of input settings for each simulation is a problem of experimental design. Sacks *et al.* identify several design objectives, but concentrate on the prediction of the response at an untried point. By viewing the deterministic response which is to be approximated as a realization of a stochastic process, they provide a statistical basis for the design of experiments.

References

Abramowitz, M. and Stegun, I. A. (1965). *Handbook of mathematical functions*. Dover Publications, New York.

Atkinson, A. C. (1972). Planning experiments to detect indaequate regression models. *Biometrika* **59**, 275–93.

Atkinson, A. C. (1973). Multifactor second order designs for cuboidal regions. *Biometrika* **60**, 15–19.

Atkinson, A. C. (1975). Planning experiments for model testing and discrimination. *Math. Oper. Statist.* **6**, 253–67.

Atkinson, A. C. (1978). Posterior probabilities for choosing a regression model. *Biometrika* **65**, 39–48.

Atkinson, A. C. (1982a). Developments in the design of experiments. *Int. Statist. Rev.* **50**, 161–77.

Atkinson, A. C. (1982b). Optimum biased coin designs for sequential clinical trials with prognostic factors. *Biometrika* **69**, 61–7.

Atkinson, A. C. (1985). *Plots, transformations, and regression: An introduction to graphical methods of diagnostic regression analysis*. Clarendon Press, Oxford.

Atkinson, A. C. (1988). Recent developments in the methods of optimum and related experimental designs. *Int. Statist. Rev.* **56**, 99–115.

Atkinson, A. C. and Cox, D. R. (1974). Planning experiments for discriminating between models (with discussion). *J. R. Statist. Soc. B* **36**, 321–48.

Atkinson, A. C. and Donev, A. N. (1989). The construction of exact D-optimum experimental designs with application to blocking response surface designs. *Biometrika* **76**, 515–26.

Atkinson, A. C. and Fedorov, V. V. (1975a). The design of experiments for discriminating between two rival models. *Biometrika* **62**, 57–70.

Atkinson, A. C. and Fedorov, V. V. (1975b). Optimal design: experiments for discriminating between several models. *Biometrika* **62**, 289–303.

Atkinson, A. C., Chaloner, K., Herzberg, A. M., and Juritz, J. (1992). Optimum experimental designs for properties of a compartmental model. *Biometrics* **48**, in press.

Bailey, R. A. (1984). Quasi-complete Latin squares: construction and randomization. *J. R. Statist. Soc. B* **46**, 323–34

Bailey, R. A. (1991). Strata for randomized experiments (with discussion). *J. R. Statist. Soc. B* **53**, 27–78.

Bandemer, H. and Näther, W. (1980). *Theorie und Anwendung der optimalen Versuchsplanung: II Handbuch zur Anwendung*. Akademie Verlag, Berlin.

Bandemer, H., Bellmann, A., Jung, W., and Richter, K. (1973). *Optimale Versuchsplanung*. Akademie Verlag, Berlin.

Bandemer, H., Bellmann, A., Jung, W., Le Anh Son, Nagel, S., Näther, W., Pilz, J., and Richter, K. (1977). *Theorie und Anwendung der optimalen Versuchsplanung: I Handbuch zur Theorie*. Akademie Verlag, Berlin.

Barnard, G. A., Box, G. E. P., Cox, D. R., Seheult, A. H., and Silverman, B. W. (ed.) (1989). *Industrial quality and reliability*. Royal Society, London.

Bates, D. M. and Watts, D. G. (1988). *Nonlinear regression and its applications*. Wiley, New York.

Baumert, L., Golomb, S. W., and Hall, M. (1962). Discovery of an Hadamard matrix of order 92. *Am. Math. Soc. Bull.* **68**, 237–8.

Becker, N. G. (1968). Models for the response of a mixture. *J. R. Statist. Soc. B* **30**, 349–58.

Becker, N. G. (1969). Regression problems when the predictor variables are proportions. *J. R. Statist. Soc. B* **31**, 107–12.

Becker, N. G. (1970). Mixture designs for a model linear in the proportions. *Biometrika* **57**, 329–38.

Begg, C. B. (1990). On inference from Wei's biased coin design for clinical trials (with discussion). *Biometrika* **77**, 467–84.

Bendell, A., Disney, J., and Pridmore, W. A. (1989). *Taguchi methods: applications in world industry*. IFS Publications, Bedford, UK.

Biggs, N. L. (1989). *Discrete mathematics* (revised edn). Clarendon Press, Oxford.

Bohachevsky, I. O., Johnson, M. E., and Stein, M. L. (1986). Generalized simulated annealing for function optimization. *Technometrics* **28**, 209–17.

Box, G. E. P. and Cox, D. R. (1964). An analysis of transformations (with discussion). *J. R. Statist. Soc. B* **26**, 211–46.

Box, G. E. P. and Draper, N. R. (1963). The choice of a second order rotatable design. *Biometrika* **50**, 335–52.

Box, G. E. P. and Draper, N. R. (1975). Robust designs. *Biometrika* **62**, 347–52.

Box, G. E. P. and Draper, N. R. (1987). *Empirical model building and response surfaces*. Wiley, New York.

Box, G. E. P. and Lucas, H. L. (1959). Design of experiments in nonlinear situations. *Biometrika* **46**, 77–90.

Box, G. E. P. and Wilson, K. B. (1951). On the experimental attainment of optimum conditions (with discussion). *J. R. Statist. Soc. B* **13**, 1–45.

Box, G. E. P., Hunter, J. S., and Hunter, W. G. (1978). *Statistics for experimenters*. Wiley, New York.

Box, G., Bisgaard, S., and Fung, C. (1989). An explanation and critique of Taguchi's contribution to quality engineering. In *Taguchi methods: Applications in world industry* (ed. A. Bendel, J. Disney, and W. A. Pridmore), pp. 359–83. IFS Publications, Bedford, UK.

Box, M. J. and Draper, N. R. (1971). Factorial designs, the $|F'F|$ criterion and some related matters. *Technometrics* **13**, 731–42. Correction, **14**, 511 (1972); **15**, 430 (1973).

Brownlee, K. A. (1965). *Statistical theory and methodology in science and engineering* (2nd edn). Wiley, New York.

Bunke, H. and Bunke, O. (ed.) (1986). *Statistical inference in linear models*, vol. 1. Wiley, New York.

Carroll, R. J. and Ruppert, D. (1988). *Transformation and weighting in regression*. Chapman & Hall, London.

Chaloner, K. (1988). An approach to experimental design for generalized linear models. In *Model-oriented data analysis* (ed. V. Fedorov and H. Läuter). Springer, Berlin.

Chaloner, K. and Larntz, K. (1989). Optimal Bayesian design applied to logistic regression experiments. *J. Statist. Planning Inf.* **21**, 191–208.

Chambers, E. A. and Cox, D. R. (1967). Discrimination between alternative binary response models. *Biometrika* **54**, 573–8.

Chan, L. Y. (1990). An example of construction of asymptotic *D*-optimal design which converges on an infinite set of points. *J. Statist. Plan. Inf.* **25**, 29–34.

Claringbold, P. J. (1955). Use of the simplex-design in the study of joint action of related hormones. *Biometrics* **11**, 174–85.

Cochran, W. G. and Cox, G. M. (1957). *Experimental designs* (2nd edn). Wiley, New York.

Connor, W. S. and Zelen, M. (1959). *Fractional factorial experiment designs for factors at three levels.* Washington, DC: National Bureau of Standards, Applied Mathematics Series No. 54.

Cook, R. D. and Nachtsheim, C. J. (1980). A comparison of algorithms for constructing exact *D*-optimum designs. *Technometrics* **22**, 315–24.

Cook, R. D. and Nachtsheim, C. J. (1982). Model robust, linear-optimal designs. *Technometrics* **24**, 49–54.

Cook, R. D. and Nachtsheim, C. J. (1989). Computer-aided blocking of factorial and response surface designs. *Technometrics* **31**, 339–46.

Cook, R. D. and Weisberg, S. (1990). Confidence curves in nonlinear regression. *J. Am. Statist. Assoc.* **85**, 544–51.

Cornell, J. A. (1990). *Experiments with mixtures* (2nd edn). Wiley, New York.

Cox, D. R. (1958). *Planning of experiments.* Wiley, New York.

Cox, D. R. (1971). A note on polynomial response functions for mixtures. *Biometrika* **58**, 155–9.

Cox, D. R. (1982). A remark on randomization in clinical trials. *Utilitas Math. A* **21**, 245–52.

Cox, D. R. and Snell, E. J. (1981). *Applied statistics.* Chapman & Hall, London.

Cox, D. R. and Snell, E. J. (1989). *Analysis of binary data* (2nd edn). Chapman & Hall, London.

Crosier, R. B. (1984). Mixture experiments: geometry and pseudocomponents. *Technometrics* **26**, 209–16.

Crowder, M. J. and Hand, D. J. (1990). Analysis of repeated measures. Chapman & Hall, London.

Daniel, C. (1959). Use of half-normal plots in interpreting factorial two-level experiments. *Technometrics* **1**, 311–41.

Davies, O. L. (ed.) (1956). *The design and analysis of industrial experiments.* Oliver and Boyd, London.

Davison, A. C., Hinkley, D. V., and Schechtman, E. (1986). Efficient bootstrap simulation. *Biometrika* **73**, 555–66.

Dehnad, K. (ed.) (1989). *Quality control, robust design, and the Taguchi method.* Wadsworth & Brooks/Cole, Pacific Grove, CA.

Derringer, G. C. (1974). An empirical model for viscosity of filled and plasticized elastomer compounds. *J. Appl. Polym. Sci.* **18**, 1083–1101.

Dobson, A. J. (1990). *An introduction to generalized linear models.* Chapman & Hall, London.

Donev, A. N. (1988). *The construction of exact D-optimum experimental designs.* Ph.D. Thesis, University of London.

Donev, A. N. (1989). Design of experiments with both mixture and qualitative factors. *J. R. Statist. Soc. B* **50**, 297–302.

Donev, A. N. and Atkinson, A. C. (1988). An adjustment algorithm for the construction of exact D-optimum experimental designs. *Technometrics* **30**, 429–33.

Draper, N. R. and St. John, R. C. (1977). A mixture model with inverse terms. *Technometrics* **19**, 37–46.

Draper, N. R. and Smith, H. (1981). *Applied regression analysis* (2nd edn). Wiley, New York.

Dubov, E. L. (1971). D-optimal designs for nonlinear models under the Bayesian approach. In *Regression experiments* (ed. V. V. Fedorov), (in Russian). Moscow University Press, Moscow.

Eccleston, J. A. and Jones, B. (1980). Exchange and interchange procedures to search for optimal designs. *J. R. Statist. Soc. B* **42**, 238–43.

Efron, B. (1971). Forcing a sequential experiment to be balanced. *Biometrika* **58**, 403–17.

Ermakov, S. M. (ed.) (1983). *The mathematical theory of planning experiments*. Nauka, Moscow.

Ermakov, S. M. and Zhigliavsky, A. A. (1987). *Mathematical theory of optimum experiments*. (in Russian). Nauka, Moscow.

Farrell, R. H., Kiefer, J., and Walbran, A. (1967). Optimum multivariate designs. In *Proc. 5th Berkeley Symp.*, Vol. 1, pp. 113–38. University of Calfornia Press, Berkeley, CA.

Fedorov, V. V. (1972). *Theory of optimal experiments*. Academic Press, New York.

Fedorov, V. V. (1981). Active regression experiments. In *Mathematical methods of experimental design* (ed. V. B. Penenko), (in Russian). Nauka, Novosibirsk.

Fedorov, V. V. and Atkinson, A. C. (1988). The optimum design of experiments in the presence of uncontrolled variability and prior information. In *Optimal design and analysis of experiments* (ed. Y. Dodge, V. V. Fedorov, and H. P. Wynn). North-Holland, Amsterdam.

Fedorov, V. and Khabarov, V. (1986). Duality of optimal designs for model discrimination and parameter estimation. *Biometrika* **73**, 183–90.

Finney, D. J. (1971). *Probit analysis* (3rd edn). Cambridge University Press, Cambridge.

Ford, I., Titterington, D. M., and Kitsos, C. (1989). Recent advances in nonlinear experimental design. *Technometrics* **31**, 49–60.

Fresen, J. (1984). *Aspects of bioavailability studies*. M.Sc. Thesis, Department of Mathematical Statistics, University of Cape Town.

Galil, Z. and Kiefer, J. (1980). Time- and space-saving computer methods, related to Mitchell's DETMAX, for finding D-optimum designs. *Technometrics* **22**, 301–13.

Galil, Z. and Kiefer, J. (1982). Construction methods for D-optimum weighing designs when $n = 3 \pmod 4$. *Ann. Statist.* **10**, 502–10.

Garvanska, P., Lekova, V., Donev, A. N., and Decheva, R. (1992). *Mikrowellenabsorbierende, electrisch zeitende Polymerpigmente für Textilmaterialien: Teil I. Optimierung der Herstellung von mikrowellenabsorbierenden Polymerpigmenten*. Submitted for publication.

Ghosh, S. (ed.) (1990). *Statistical design and analysis of industrial experiments*. Dekker, New York.

Gittins, J. C. (1989). *Multi-armed bandit allocation indices*. Wiley, Chichester.

Graham, R. L., Hinkley, D. V., John, P. W. M., and Shi, S. (1990). Balanced design of bootstrap simulations. *J. R. Statist. Soc. B* **52**, 185–202.

Guest, P. G. (1958). The spacing of observations in polynomial regression. *Ann. Math. Statist.* **29**, 294–99.

Haines, L. M. (1987). The application of the annealing algorithm to the construction of exact *D*-optimum designs for linear-regression models. *Technometrics* **29**, 439–47.

Harville, D. (1974). Nearly optimal allocation of experimental units using observed covariate values. *Technometrics* **16**, 589–99.

Harville, D. (1975). Computing optimum designs for covariate models. In *A survey of statistical design and linear models* (ed. J. N. Srivastava), pp. 209–28. North-Holland, Amsterdam.

Heckman, N. E. (1985). A local limit theorem for a biased coin design for sequential tests. *Ann. Statist.* **13**, 209–28.

Herzberg, A. M. and Cox, D. R. (1969). Recent work on the design of experiments: a bibliography and a review. *J. R. Statist. Soc. A* **132**, 29–67.

Hines, W. L. and Montgomery, D. C. (1990). *Probability and statistics in engineering and management science* (3rd ed). Wiley, New York.

John, J. A. and Quenouille, M. H. (1977). *Experiments: Design and analysis.* Griffin, London.

John, P. W. M. (1971). *Statistical design and analysis of experiments.* Macmillan, New York.

Johnson, M. E. and Nachtsheim, C. J. (1983). *D*-optimal design on convex design spaces. *Technometrics* **25**, 271–77.

Jones, B. (1976). An algorithm for deriving optimal block designs. *Technometrics* **18**, 451–8.

Jones, B. and Eccleston, J. A. (1980). Exchange and interchanges procedures to search for optimal designs. *J. R. Statist. Soc. B* **42**, 238–43.

Juusola, J. A., Bacon, D. W., and Downie, J. (1972). Sequential statistical design strategy in an experimental kinetic study. *Can. J. Chem. Eng.* **50**, 796–801.

Kennard, R. W. and Stone, L. (1969). Computer aided design of experiments. *Technometrics* **11**, 137–48.

Kenworthy, O. O. (1963). Factorial experiments with mixtures using ratios. *Ind. Quality Control* **19**, 24–6.

Khuri, A. I. (1984). A note on *D*-optimal designs for partially nonlinear regression models. *Technometrics* **26**, 59–61.

Kiefer, J. (1959). Optimum experimental designs (with discussion). *J. R. Statist. Soc. B* **21**, 272–319.

Kiefer, J. (1961). Optimum designs in regression problems II. *Ann. Math. Statist.* **32**, 298–325.

Kiefer, J. (1974). General equivalence theory for optimum designs (approximate theory). *Ann. Statist.* **2**, 849–79.

Kiefer, J. (1975). Optimal design: variation in structure and performance under change of criterion. *Biometrika* **62**, 277–88.

Kiefer, J. (1985). *Jack Carl Kiefer collected papers*, (ed. L. D. Brown, I. Olkin, J. Sacks, H. P. Wynn). Volume III (Design of Experiments). Springer-Verlag, New York.

Kiefer, J. and Wolfowitz, J. (1959). Optimum designs in regression problems. *Ann. Math. Statist.* **30**, 271–94.

Kiefer, J. and Wolfowitz, J. (1960). The equivalence of two extremum problems. *Can. J. Math.* **12**, 363–6.

Kiefer, J. and Wynn, H. P. (1984). Optimum and minimax exact treatment designs for one-dimensional autoregressive error process. *Ann. Statist.* **12**, 431–50.

Kitsos, C. P., Titterington, D. M., and Torsney, B. (1988). An optimal design problem in rhythmometry. *Biometrics* **44**, 657–71.

Kôno, K. (1962). Optimum designs for quadratic regression on *k*-cube. *Mem. Fac. Sci. Fyushu Univ. A* **16**, 114–22.

Kurotschka, V. G. (1981). A general approach to optimum design of experiments with qualitative and quantitative factors. In *Proc. Indian Statistical Institute Golden Jubilee Int. Conf. on Statistics: Applications and new directions* (ed. J. K. Ghosh and J. Roy), pp. 353–68. Indian Statistical Institute, Calcutta.

Kuroturi, I. S. (1966). Experiments with mixtures of components having lower bounds. *Ind. Quality Control* **22**, 592–6.

Läuter, E. (1974). Experimental design in a class of models. *Math. Oper. Statist.* **5**, 379–96.

Läuter, E. (1976). Optimal multipurpose designs for regression models. *Math. Oper. Statist.* **7**, 51–68.

Lim, Y. B., Studden, W. J., and Wynn, H. P. (1988). A note on approximate *D*-optimal designs for $G \times 2^m$. In *Statistical decision theory and related topics IV* (ed. J. O. Burger and S. S. Gupta), Vol. 2, pp. 351–61. New York: Springer.

Logothetis, N. and Wynn, H. P. (1989). *Quality through design.* Clarendon Press, Oxford.

Lucas, J. M. (1976). Which response surface design is best. *Technometrics* **18**, 411–17.

Martin, R. J. (1986). On the design of experiments under spatial correlation. *Biometrika* **73**, 247–77.

McCullagh, P. and Nelder, J. A. (1989). *Generalized linear models.* (2nd edn). Chapman & Hall, London.

McLean, R. A. and Anderson, V. L. (1966). Extreme vertices design of mixture experiments. *Technometrics* **8**, 447–54.

Mead, R. (1988). *The design of experiments.* Cambridge University Press, Cambridge.

Miller, A. J. (1990). *Subsets selection in regression.* Chapman & Hall, London.

Mitchell, T. J. (1974a). An algorithm for the construction of '*D*-optimum' experimental designs. *Technometrics* **16**, 203–10.

Mitchell, T. J. (1974b). Computer construction of '*D*-optimal' first order designs. *Technometrics* **16**, 211–20.

Mitchell, T. J. and Bayne, C. K. (1978). *D*-optimal fractions of three-level factorial designs. *Technometrics* **20,** 369–80.

Mitchell, T. J. and Miller, F. L. (1970). Use of design repair to construct designs for special linear models. *Rep. ORNL–4661*, pp. 130–1. Oak Ridge National Laboratory, Oak Ridge, TN.

Nachtsheim, C. J. (1989). On the design of experiments in the presence of fixed covariates. *J. Statist. Plan. Inf.* **22**, 203–12.

Nalimov, V. V. (ed.) (1982) *Tables for planning experiments for factorials and polynomial models.* (in Russian), Metallurgica, Moscow.

Newbold, P. (1988). *Statistics for business and economics* (2nd edn). Prentice-Hall, Englewood Cliffs, NJ.

Pazman, A. (1986). *Foundations of optimum experimental design.* Reidel, Dordrecht.

Pearce, S. C. (1983). *The agricultural field experiment.* Interscience, Chichester.

Pesotchinsky, L. L. (1975). *D*-optimum and quasi-*D*-optimum second order designs on a cube. *Biometrika* **62**, 335–45.

Petkova, E., Shkodrova, V., Vassilev, H., Donev, A. N. (1987). Optimization of the purfication of nickel sulphate solution in the joint presence of iron (II), copper (II), and zinc (II). *Metallurgia* **6**, 12–17 (in Bulgarian).

Piepel, G. F. and Cornell, J. A. (1985). Models for mixture experiments when the response depends on the total amount. *Technometrics* **27**, 219–27.

Pilz, J. (1983). *Bayesian estimation and design in linear models*. Teubner, Leipzig.

Pilz, J. (1989). *Bayesian estimation and experimental design in linear regression models*. Wiley, New York.

Plackett, R. L. and Burman, J. P. (1946). The design of optimum multifactorial experiments. *Biometrika* **33**, 305–25.

Ponce de Leon, A. C. (1992). *Optimum experimental designs for binary data models*. Ph.D. Thesis, University of London.

Ponce de Leon, A. C. and Atkinson, A. C. (1991). Optimum experimental design for discriminating between two rival models in the presence of prior information. *Biometrika* **78**, 601–8.

Pronzato, L. and Walter, E. (1985). Robust experimental design via stochastic approximation. *Math. Biosci.* **75**, 103–20.

Rafajlowicz, E. (1989). Time-domain optimization of input signals for distributed-parameter systems identification. *J. Optim. Theor. Appl.* **60**, 67–79.

Rao, C. R. (1973). *Linear statistical inference and its applications* (2nd edn). Wiley, New York.

Rasch, D. and Herrendörfer, G. (1982). *Statistische Versuchsplanung*. Deutscher Verlag der Wissenschaften, Berlin.

Ratkowsky, D. A. (1983). *Nonlinear regression modeling*. Dekker, New York.

Ratkowsky, D. A. (1990). *Handbook of nonlinear regression models*. Dekker, New York.

Ripley, B. D. (1987). *Stochastic simulation*. Wiley, New York.

Sacks, J., Welch, W. J., Mitchell, T. J., and Wynn, H. P. (1989). Design and analysis of computer experiments. *Statist. Sci.* **4**, 409–35.

Sams, D. A. and Shadman, F. (1986). Mechanism of potassium-catalyzed carbon/CO_2 reaction. *AIChE J.* **32**, 1132–7.

Savova, I., Donev, T. N., Tepavicharova, I., and Alexandrova, T. (1989). Comparative studies on the storage of freeze-dried yeast strains on the genus Saccharomyces. In *Proc 4th Int. School on Cryobiology and Freeze-drying, 29 July–6 August 1989, Borovets Bulgaria*, pp. 32–3. Bulgarian Academy of Sciences Press, Sofia.

Saxena, S. K. and Nigam, A. K. (1973). Symmetric-simplex block designs for mixtures. *J. R. Statist. Soc. B* **35**, 466–72.

Scheffé, H. (1958). Experiments with mixtures. *J. R. Statist. Soc. B* **20**, 344–60.

Seber, G. A. F. (1977). *Linear regression analysis*. Wiley, New York.

Seber, G. A. F. and Wild, C. J. (1989). *Nonlinear regression*. Wiley, New York.

Shah, K. R. and Sinha, B. K. (1989). *Theory of optimal design*. Lecture Notes in Statistics 54. Springer, Berlin.

Shelton, J. T., Khuri, A. I., and Cornell, J. A. (1983). Selecting check points for testing lack of fit in response surface models. *Technometrics* **25**, 357–65.

Sibson, R. (1974). D_A-optimality and duality. In *Progress in Statistics*, Vol. 2, Proc. 9th European Meeting of Statisticians, Budapest, 1972. (ed. J. Gani, K. Sarkadi, and I. Vincze). North-Holland, Amsterdam.

Silvey, S. D. (1980). *Optimum design*. Chapman & Hall, London.

Silvey, S. D., Titterington, D. M., and Torsney, B. (1978). An algorithm for optimal designs on a finite design space. *Commun. Statist. A* **7**, 1379–89.

Smith, A. F. M. and Spiegelhalter, D. J. (1980). Bayes factors and choice criteria for linear models. *J. R. Statist. Soc. B* **42**, 213–20.

Smith, K. (1918). On the standard deviations of adjusted and interpolated values of an observed polynomial function and its constraints and the guidance they give towards a proper choice of the distribution of observations. *Biometrika* **12**, 1–85.

Smith, R. L. (1984*a*). Properties of biased coin designs in sequential clinical trials. *Ann. Statist.* **12**, 1018–34.

Smith, R. L. (1984*b*). Sequential treatment allocation using biased coin designs. *J. R. Statist. Soc. B* **46**, 519–43.

Snee, R. D. (1985). Computer-aided design of experiments—some practical experiences. *J. Qual. Technol.* **17**, 222–36.

Spezzaferri, F. (1988). Nonsequential designs for model discrimination and parameter estimation. In *Bayesian statistics* 3 (ed. J. M. Bernardo, M. H. DeGroot, D. V. Lindley, and A. F. M. Smith) pp. 777–83. Clarendon Press, Oxford.

Spiegelhalter, D. J. and Smith, A. F. M. (1982). Bayes factors for linear and log-linear models with vague prior information. *J. R. Statist. Soc. B* **44**, 377–87.

Street, A. P. and Street, D. J. (1987). *Combinatorics of experimental design.* Clarendon Press, Oxford.

Street, D. J. and Street, A. P. (1985). Designs with partial neighbour balance. *J. Statist. Plan. Inf.* **12**, 47–59.

Taguchi, G. (1987). *Systems of experimental design* (Vols 1 and 2, 1976 and 1977, with 1987 translation). UNIPUB, Langham, MD.

Titterington, D. M. (1980). Aspects of optimal design in dynamic systems. *Technometrics* **22**, 287–99.

Van Schalkwyk, D. J. (1971). *On the design of mixture experiments.* Ph.D. Thesis, University of London.

Vuchkov, I. N. (1977). A ridge-type procedure for design of experiments. *Biometrika* **64**, 147–50.

Vuchkov, I. N. (1982). Sequentially generated designs. *Biometric J.* **24**, 751–63.

Vuchkov, I. N., Yontchev, C. A., Damgaliev, D. L., Tsochev, V. V., and Dikova, T. D. (1978). *Catalogue of sequentially generated designs.* Higher Institute of Chemical Technology Press, Sofia.

Vuchkov, I. N., Damgaliev, D. L., and Yontchev, C. A. (1981). Sequentially generated second order quasi *D*-optimal designs for experiments with mixture and process variables. *Technometrics* **23**, 233–8.

Wei, L.-J. (1978*a*). The adaptive biased-coin design for sequential experiments. *Ann. Statist.* **6**, 92–100.

Wei. L.-J. (1978*b*). An application of an urn model to the design of sequential controlled clinical trials. *J. Am. Statist. Assoc.* **73**, 559–63.

Wei, L.-J., Smythe, R. T., and Smith, R. L. (1986). K-treatment comparisons with restricted randomization rules in clinical trials. *Ann. Statist.* **14**, 265–74.

Weisberg, S. (1985). *Applied linear regression* (2nd edn). Wiley, New York.

Welch, W. J. (1982). Branch and bound search for experimental designs based on *D*-optimality and other criteria. *Technometrics* **24**, 41–8.

Welch, W. J. (1984). Computer-aided design of experiments for response estimation. *Technometrics* **26**, 217–24.

Whittle, P. (1973). Some general points in the theory of optimal experimental design. *J. R. Statist. Soc. B* **35**, 123–30.

Wierich, W. (1986). On optimal designs and complete class theorems for experiments with continuous and discrete factors of influence. *J. Statist. Plan. Inf.* **15**, 19–27.

Williams, E. R. (1985). A criterion for the construction of optimal neighbour designs. *J. R. Statist. Soc. B* **47**, 489–97.

Wilson, E. B. (1952). *An introduction to scientific research.* McGraw-Hill, New York.

Wu, C. F. J. and Wynn, H. P. (1978). The convergence of general step length algorithms for regular optimum design criteria. *Ann. Statist.* **6**, 1286–1301.

Wynn, H. P. (1970). The sequential generation of *D*-optimum experimental designs. *Ann. Math. Statist.* **41**, 1655–64.

Wynn, H. P. (1972). Results in the theory and construction of *D*-optimum experimental designs. *J. R. Statist. Soc. B* **34**, 133–47.

Wynn, H. P. (1984). Jack Kiefer's contributions to experimental design. *Ann. Statist.* **12**, 416–23.

Wynn, H. P. and Ogbonmwan, S. M. (1986). Discussion of 'Jacknife, bootstrap and other resampling methods in regression analysis' by C. F. J. Wu. *Ann. Statist.* **14**, 1340–3.

Zarrop, M. B. (1979). *Optimal experimental design for dynamic system identification.* Lecture Notes in Control and Information Sciences 21. Springer, New York.

Appendix A
Program to implement the BLKL algorithm of §15.6

The program is written in Fortran 77. The subroutine which implements the BLKL algorithm is called BLKL. All necessary subroutines are also listed except the following.

RAND (Z) which generates a uniformly distributed random number in the interval [0, 1].

MINV (W, P, PP, DET, IR, IV) which inverts a $PP = P \times P$ vector matrix W and calculates the determinant DET; IR and IV are working vectors of size P.

Description of the input to the program:

Identifier	Meaning	Restriction
M	Total number of factors	1–4
MM	Number of mixture components	3–4
NBL	Number of blocks	1–4
N	Number of trials	P–50
NI (NBL)	Block sizes	1–N–NBL
LL	Order of the model	1–4
P	Number of model parameters	2–19
LB (LL)	Indices of the model parameters	
K	Number of design points considered for deletion	1–N
L	Number of design points considered for inclusion	1–support
IE	Number of tries	
J1	Number of trials to be included in the design	0–N
X (J1, M + NBL)	Trials to be included in the design	
LSU	Support to be read (0 – Yes; 1 – No)	
LSUP	Number of support points to be read	
X (LSUP, M + NBL)	Support points (if LSUP > 0).	1–324
IW	Extending listing (0 – Yes; 1 – No)	

Alteration of these constants is possible and depends on which Fortran version is used.

```
        DIMENSION W(361),LB(19,4),X(325,8),F(325,19),IV(19),IR(19)
        DIMENSION NI(4),STEP(8),LEVELS(8),D(325),NSUP(325),NDES(50)
        DIMENSION XX(19),NOPT(50),NIO(4)
        INTEGER P,PP,H
        OPEN(UNIT=1,FILE='IN.DAT',STATUS='OLD')
        OPEN(UNIT=3,FILE='OUT.DAT',STATUS='UNKNOWN')
1       FORMAT(6X,C(1X,F6.3))
2       FORMAT(I6,6(1X,F6.3))
3       FORMAT(///,3X,'The design:',/)
4       FORMAT(/,' M=',I3,' P=',I3,' N=',I3,' Order',I3,' Tries=',
      * I3,/)
5       FORMAT(' MM=',I3,' J1=',I3,' K=',I3,' L=',I3,' NBlocks=',
      * I3,/)
6       FORMAT(' Block sizes: ',5I5,/)
7       FORMAT(/,' Det =',E15.5,' DetN=',E15.5,/)
8       FORMAT(//,' Maximum Det =',E15.5,/,' Maximum DetN=',
      * E15.5,/)
9       FORMAT(//,3X,'The design after the adjustment algorithm:',/)
10      FORMAT(10I4)
11      FORMAT(//,'Try',I4,' Start Det ',E15.5,/)
12      FORMAT(5X,'Try ',I3)
13      FORMAT(//,'Model structure',/)
14      FORMAT(21I3)
15      FORMAT(//)
16      FORMAT(//,' Dave=',E12.5,' Dmax=',E12.5)
        READ(1,*)M
        READ(1,*)MM
        READ(1,*)NBL
        READ(1,*)N
        IF(NBL.GT.1)GOTO 21
        NI(1)=N
        MNBL=M
        GOTO 22
21      READ(1,*) (NI(I),I=1,NBL)
        MNBL=M+NBL
22      READ(1,*)LL
        READ(1,*)P
        DO 23 I=1,P
23      READ(1,*) (LB(I,II),II=1,LL)
        READ(1,*)K
        READ(1,*)L
        READ(1,*)IE
        READ(1,*)J1
```

```
          IH=0
          IF(J1.EQ.0)GOTO 25
          DO 24 I=1,J1
          NDES(I)=I
          NSUP(I)=I
          NOPT(I)=I
          IH=IH+1
24        READ(1,*) (X(I,II),II=1,MNBL)
25        READ(1,*)LSUP
          IF(LSUP.EQ.0)READ(1,*)LSUP
          WRITE(3,4)M,P,N,LL,IE
          WRITE(3,5)MM,J1,K,L,NBL
          IF(NBL.GT.1)WRITE(3,6) (NI(I),I=1,NBL)
          WRITE(3,13)
          DO 26 I=1,LL
26        WRITE(3,14) (LB(II,I),II=1,P)
          WRITE(3,15)
          DETMAX=0.
          PP=P*P
          IF(N.EQ.J1)GOTO 38
          REG=0.001
          REGI=1./REG
          Z=0.1241956
          IF(LSUP.GT.1)GOTO 36
          H=1
          DO 29 I=1,M
          JJ=0
          DO 28 II=1,P
          J=0
          DO 27 III=1,LL
          IF(LB(II,III).EQ.I)J=J+1
27        CONTINUE
          IF(J.GT.JJ)JJ=J
28        CONTINUE
          LEVELS(I)=JJ+1
          IF(I.LE.MM)LEVELS(I)=LEVELS(I)+1
          H=H*LEVELS(I)
29        STEP(I)=2./(LEVELS(I)-1)
          H=H*NBL
          DO 35 I=1,H,NBL
          CALL MPT(XX,I,M,STEP)
          IF(MM.EQ.0)GOTO31
          A=0.
```

```
      DO 30 II=1,MM
30    A=A+(XX(II)+1.)/2.
      IF(A.LT.0.98.OR.A.GT.1.02)GOTO 35
31    DO 34 II=1,NBL
      IH=IH+1
      JM=II+M
      DO 32 III=1,NBL
32    XX(III+M)=0.
      XX(JM)=1.
      NSUP(IH)=IH
      DO 33 III=1,MNBL
      A=XX(III)
      IF(III.LE.MM)A=(A+1.)/2.
33    X(IH,III)=A
34    CONTINUE
35    CONTINUE
      GOTO 38
36    DO 37 I=1,LSUP
      IH=IH+1
      NSUP(IH)=IH
37    READ(1,*) (X(IH,II),II=1,MNBL)
38    H=IH
      CALL FMATRI(H,P,LL,LB,X,F)
      READ(1,*)IW
      IF(N.EQ.J1)GOTO 45
      DO 44 ITRIES=1,IE
      DET=1.
      DO 39 I=1,NBL
39    NIO(I)=NI(I)
      IF(J1.EQ.0)THEN
      J=0
      DO 40 I=1,P-1
      J=J+1
      W(J)=REGI
      DO 40 II=1,P
      J=J+1
40    W(J)=0.
      W(PP)=REGI
      ELSE
      CALL GMATRI(J1,P,PP,F,W,NDES)
      DO 41 I=1,P
      II=I+(I-1)*P
41    W(II)=W(II)+REG
```

```
          CALL MINV(W,P,PP,DET,IR,IV)
          END IF
          CALL VAR(P,D,F,W,H)
          CALL STAGE1(W,F,H,P,DET,D,NDES,NSUP,Z,J1,N,NIO,NBL)
          CALL GMATRI(N,P,PP,F,W,NDES)
          CALL MINV(W,P,PP,DET,IR,IV)
          IF(IW.EQ.1)WRITE(3,11)ITRIES,DET
          IF(DET.LE.1.E−32)GOTO 44
          CALL VAR(P,D,F,W,H)
          CALL BLKL(D,W,F,H,P,PP,DET,N,NDES,NSUP,J1,K,L,NBL,IV,IR,IW)
          CALL GMATRI(N,P,PP,F,W,NDES)
          CALL MINV(W,P,PP,DET,IR,IV)
          IF(IW.EQ.1)WRITE(3,10) (NDES(I),I=1,N)
          DETN=DET
          DO 42 I=1,P
42        DETN=DETN/N
          IF(IW.EQ.1)WRITE(3,7)DET,DETN
          WRITE(*,12)ITRIES
          IF(DET.LT.DETMAX)GOTO 44
          DETMAX=DET
          DO 43 I=1,N
43        NOPT(I)=NDES(I)
44        CONTINUE
45        CALL GMATRI(N,P,PP,F,W,NOPT)
          CALL MINV(W,P,PP,DET,IR,IV)
          DETN=DET
          DO 46 I=1,P
46        DETN=DETN/N
          WRITE(3,8)DET,DETN
          WRITE(3,3)
          DO 47 I=1,N
          J=NOPT(I)
47        WRITE(3,2)J,(X(J,II),II=1,MNBL)
          STOP
          END
C
          SUBROUTINE FMATRI(N,P,LL,LB,X,F)
          DIMENSION LB(19,*),X(325,*),F(325,*)
          INTEGER P,PP,H
          DO 2 I=1,N
          DO 2 J=1,P
          F(I,J)=1.
          DO 2 IJ=1,LL
```

```
         IF(LB(J,IJ))2,2,1
1        F(I,J)=F(I,J)*X(I,LB(J,IJ))
2        CONTINUE
         RETURN
         END
C
         SUBROUTINE GMATRI(N,P,PP,F,W,NDES)
         DIMENSION F(325,*),W(*),NDES(*)
         INTEGER P,PP,H
         DO 1 I=1,PP
1        W(I)=0
         DO 2 I=1,N
         J=NDES(I)
         DO 2 II=1,P
         IJ=0
         DO 2 JI=II,PP,P
         IJ=IJ+1
2        W(JI)=W(JI)+F(J,IJ)*F(J,II)
         RETURN
         END
C
         SUBROUTINE VAR(P,D,F,W,H)
         DIMENSION F(325,*),W(*),D(*)
         INTEGER P,PP,H
         DO 3 I=1,H
         R=0.
         DO 2 II=1,P
         S=0.
         J=(II-1)*P
         DO 1 III=1,P
         JJ=III+J
1        S=S+W(JJ)*F(I,III)
2        R=R+F(I,II)*S
3        D(I)=R
         RETURN
         END
C
         SUBROUTINE MPT(X,I,M,STEP)
         DIMENSION STEP(*),X(*)
         INTEGER P,PP,H
         IF(I.GT.1)GOTO 2
         DO 1 J=1,M
1        X(J)=-1.
```

```
        RETURN
2       DO 3 J=1,M
        IF(X(J).GE.(1.-STEP(J)/2.))GOTO 3
        X(J)=X(J)+STEP(J)
        RETURN
3       X(J)=-1.
        RETURN
        END
C
        SUBROUTINE ORDER(D,NN,NST,NSIZE,K,NOPT)
        DIMENSION D(*),NN(*)
        INTEGER P,PP,H
        DO 3 I=1,K
        DO 3 II=1,NSIZE-I-NST+1
        J=NSIZE-II+1
        IF(NOPT.EQ.2)GOTO 1
        IF(D(NN(J)).LE.D(NN(J-1)))GOTO 3
        GOTO 2
1       IF(D(NN(J)).GE.D(NN(J-1)))GOTO 3
2       JJ=NN(J-1)
        NN(J-1)=NN(J)
        NN(J)=JJ
3       CONTINUE
        RETURN
        END
C
        SUBROUTINE ADDPT(D,W,F,H,P,DET,N,NDES,JJJ)
        DIMENSION W(*),F(325,*),WFT(19),D(*),NDES(*)
        INTEGER P,PP,H
        S=1.+D(JJJ)
        DET=DET*S
        DO 2 I=1,P
        J=(I-1)*P
        T=0.
        DO 1 II=1,P
1       T=T+W(II+J)*F(JJJ,II)
2       WFT(I)=T
        DO 3 I=1,P
        J=(I-1)*P
        SS=WFT(I)/S
        DO 3 II=1,P
        JJ=II+J
3       W(JJ)=W(JJ)-WFT(II)*SS
```

```
      DO 5 I=1,H
      T=0.
      DO 4 II=1,P
4     T=T+WFT(II)*F(I,II)
5     D(I)=D(I)-T*T/S
      N=N+1
      NDES(N)=JJJ
      RETURN
      END
C
      SUBROUTINE STAGE1(W,F,H,P,DET,D,NDES,NSUP,Z,J1,ND,NI,NBL)
      DIMENSION W(*),F(325,*),NDES(*),D(*),NSUP(*),NI(*)
      INTEGER P,PP,H
      CALL RAND(Z)
      J2=Z*(P/2)+1
      N=J1
      J=H-J1
      DO 2 I=1,J2
1     CALL RAND(Z)
      JJ=Z*J+1+J1
      IF(NBL.EQ.1)GOTO 2
      JJJ=(JJ-J1)/NBL
      JJJ=JJ-JJJ*NBL-J1
      IF(JJJ.EQ.0)JJJ=NBL
      IF(NI(JJJ).EQ.0)GOTO 1
      NI(JJJ)=NI(JJJ)-1
2     CALL ADDPT(D,W,F,H,P,DET,N,NDES,JJ)
      DO 4 I=1,ND-J1-J2
      NST=J1+1
3     CALL ORDER(D,NSUP,NST,J,1,1)
      JJ=NSUP(NST)
      IF(NBL.EQ.1)GOTO 4
      JJJ=(JJ-J1)/NBL
      JJJ=JJ-JJJ*NBL-J1
      IF(JJJ.EQ.0)JJJ=NBL
      IF(NI(JJJ).EQ.0)THEN
      NST=NST+1
      GOTO 3
      ELSE
      NI(JJJ)=NI(JJJ)-1
      END IF
4     CALL ADDPT(D,W,F,H,P,DET,N,NDES,JJ)
      RETURN
      END
```

```
C
        SUBROUTINE BLKL(D,W,F,H,P,PP,DET,N,NDES,NSUP,J1,K,L,NBL,
     *  IV,IR,IW)
        DIMENSION W(*),F(325,*),WFT(19),D(*),NDES(*),NSUP(*)
        DIMENSION IV(*),IR(*)
        INTEGER P,PP,H
        J2=J1+1
        NUM=0
1       IF(K.NE.N)CALL ORDER(D,NDES,J2,N,K,2)
        IF(L.NE.H)CALL ORDER(D,NSUP,J2,H-J1,L,1)
        DELM=1.
        DO 5 I=J2,J1+K
        JK=NDES(I)
        JJJ=JK/NBL
        JJJ=JK-JJJ*NBL
        DK=D(JK)
        DO 3 II=1,P
        A=0.
        DO 2 III=1,P
        J=(III-1)*P+II
2       A=A+F(JK,III)*W(J)
3       WFT(II)=A
        DO 5 II=J2,J1+L
        JL=NSUP(II)
        JJL=JL/NBL
        JJL=JL-JJL*NBL
        IF(JJJ.NE.JJL)GOTO 5
        DKL=0.
        DO 4 III=1,P
4       DKL=DKL+WFT(III)*F(JL,III)
        DELTA=(1.+D(JL))*(1.-DK)+DKL*DKL
        DELTA=(DELTA+P-1.)/P
        IF(DELTA.LE.DELM)GOTO 5
        DELM=DELTA
        JKK=I
        JLL=JL
5       CONTINUE
        IF(DELM.LT.1.0001)RETURN
        JJ=NDES(JKK)
        NDES(JKK)=JLL
        CALL GMATRI(N,P,PP,F,W,NDES)
        CALL MINV(W,P,PP,DETA,IR,IV)
        IF(DETA.GT.DET)GOTO 6
        NDES(JKK)=JJ
```

```
      RETURN
6     DET=DETA
      CALL VAR(P,D,F,W,H)
      NUM=NUM+1
      IF(IW.EQ.1)WRITE(3,7)NUM,DET,JLL,JJ
7     FORMAT(' Iter',I3,' Det=',E15.5,' In ',I4,' Out ',I4)
      GOTO 1
      END
```

The following four examples illustrate the use of the program. Both input and output files are listed. The design points are represented by their numbers in the list of candidate points. For designs on points of the 3^m factorial these are the numbers of the standard ordering used in Tables 11.6 and 11.7. These numbers are given for the optimum design together with the levels of the factors. Models are described by the indices of the parameters: for example 23 in the model structure means that β_{23} is included in the model.

Example A1

The construction of a 14-trial design for the second-order model in three factors is shown, for $K = L = 3$, six tries, and support the points of the 3^m factorial. A full output was required.

Input file

IN.DAT:

```
3
0
1
14
2
10
0 0
1 0
2 0
3 0
1 2
1 3
2 3
1 1
2 2
3 3
3
3
6
```

0
1
1

Output file

OUT.DAT:

M=3 P=10 N=14 Order 2 Tries=6
MM=0 J1=0 K=3 L=3 NBlocks=1

Model structure

```
0   1   2   3   1   1   2   1   2   3
0   0   0   0   2   3   3   1   2   3
```

Try   1 Start Det   0.13006E+09
```
   25  26  22   5  21  16  19   1   3   7
    9  15  11  27
```
   Det=0.13006E+09   DetN=0.44963E−03

Try   2 Start Det   0.83231E+08
```
   Iter   1   Det=0.10764E+09 In    2 Out   24
   Iter   2   Det=0.13006E+09 In   10 Out   11
    1  10   2   4  15  17  23   7   3  19
   25  21   9  27
```
   Det=0.13006E+09   DetN=0.44963E−03

Try   3 Start Det   0.11243E+09
```
   22  12  20  13  19  21   7  27   3   9
    1  25   5  17
```
   Det=0.11243E+09   DetN=0.38868E−03

Try   4 Start Det   0.56713E+08
```
   Iter   1   Det=0.72770E+08 In   21 Out   17
   21  24  22  17   6   1  10   2  12  20
   27  25   7   9
```
   Det=0.72770E+08   DetN=0.25158E−03

Try   5 Start Det   0.13107E+09
```
   13  11   5  23  21   9   3  27   7  25
   19  17  15   1
```
   Det=0.13107E+09   DetN=0.45314E−03

Try   6 Start Det   0.88412E+08
```
    9  26  18   8   1  25  21  19  27   3
    6  16  14   4
```
   Det=0.88412E+08   DetN=0.30566E−03

   Maximum Det=0.13107E+09
   Maximum DetN=0.45314E−03

The design:

13	−1.000	0.000	0.000
11	0.000	−1.000	0.000
5	0.000	0.000	−1.000
23	0.000	0.000	1.000
21	1.000	−1.000	1.000
9	1.000	1.000	−1.000
3	1.000	−1.000	−1.000
27	1.000	1.000	1.000
7	−1.000	1.000	−1.000
25	−1.000	1.000	1.000
19	−1.000	−1.000	1.000
17	0.000	1.000	0.000
15	1.000	0.000	0.000
1	−1.000	−1.000	−1.000

It is interesting to note that the algorithm obtained the composite design for three factors of the form described in §7.5.

Example A2

We now construct a 13-trial design in two blocks for a second-order model, $K = 3$, $L = 5$, 10 tries, and support the points of the 3^2 factorial repeated for each of the block variables x_3 and x_4. The block sizes were prespecified to be 5 and 8. This time we require a short output.

Input file

IN.DAT:

```
2
0
2
13
5 8
2
7
1 0
2 0
3 0
4 0
1 2
1 1
2 2
3
5
```

10
0
1
0

Output file

OUT.DAT:
M=2 P=7 N=13 Order 2 Tries=10
MM=0 J1=0 K=3 L=5 NBlocks=2
Block sizes: 5   8

Model structure
    1   2   3   4   1   1   2
    0   0   0   0   2   1   2

Maximum Det=0.16640E+06
Maximum DetN=0.26519E−02

The design:
     2 −1.000 −1.000   0.000   1.000
    14 −1.000   1.000   0.000   1.000
    18   1.000   1.000   0.000   1.000
    12   1.000   0.000   0.000   1.000
     6   1.000 −1.000   0.000   1.000
     8 −1.000   0.000   0.000   1.000
     4   0.000 −1.000   0.000   1.000
    16   0.000   1.000   0.000   1.000
    17   1.000   1.000   1.000   0.000
     5   1.000 −1.000   1.000   0.000
     1 −1.000 −1.000   1.000   0.000
    13 −1.000   1.000   1.000   0.000
     9   0.000   0.000   1.000   0.000

The design obtained is the one given in Fig. 14.2(b).

Example A3

This is an example of the augmentation of a nine-trial second-order design to a 13-trial third-order design. The resulting design is shown in Fig. 17.2. Design points 1–9 are those of the design for augmentation.

Input file

IN.DAT:
2
0
1

```
13
3
10
0 0 0
1 0 0
2 0 0
1 2 0
1 1 0
2 2 0
1 1 2
1 2 2
1 1 1
2 2 2
4
5
10
9
−1.000 −1.000
 0.0000 −1.000
 1.0000 −1.000
−1.000  0.0000
 0.0000  0.0000
 1.0000  0.0000
−1.000  1.0000
 0.0000  1.0000
 1.0000  1.0000
1
0
```

Output file

OUT.DAT:

M=2 P=10 N=13 Order 3 Tries=10
MM=0 J1=9 K=5 L=5 NBlocks=1

Model structure

```
0   1   2   1   1   2   1   1   1   2
0   0   0   2   1   2   1   2   1   2
0   0   0   0   0   0   2   2   1   2
```

Maximum Det=0.30823E+04
Maximum DetN=0.22358E−07

The design:
```
    1 −1.000 −1.000
    2  0.000 −1.000
```

```
3   1.000 −1.000
4 −1.000   0.000
5   0.000   0.000
6   1.000   0.000
7 −1.000   1.000
8   0.000   1.000
9   1.000   1.000
15 −0.333 −0.333
16   0.333 −0.333
19 −0.333   0.333
20   0.333   0.333
```

Example A4

We now show how a 10-trial design for a mixture experiment with non-standard design region is constructed. The design is given in Fig. 16.3.

Input file

IN.DAT:

```
3
3
1
10
2
6
1 0
2 0
3 0
1 2
1 3
2 3
3
5
5
0
0
13
0.7000    0.1000    0.2000
0.2000    0.6000    0.2000
0.7000    0.2000    0.1000
0.2000    0.2000    0.6000
0.3000    0.6000    0.1000
0.3000    0.1000    0.6000
```

```
0.7000    0.1500    0.1500
0.2000    0.4000    0.4000
0.5000    0.1000    0.4000
0.2500    0.6000    0.1500
0.5000    0.4000    0.1000
0.2500    0.1500    0.6000
0.4000    0.3000    0.3000
0
```

Output file

OUT.DAT:
M=3 P=6 N=10 Order 2 Tries=5
MM=3 J1=0 K=3 L=5 NBlocks=1

Model structure
```
  1   2   3   1   1   2
  0   0   0   2   3   3
```

Maximum Det=0.79702E−07
Maximum DetN=0.797023E−13

The design:
```
  13   0.400   0.300   0.300
   8   0.200   0.400   0.400
   9   0.500   0.100   0.400
  11   0.500   0.400   0.100
   3   0.700   0.200   0.100
   2   0.200   0.600   0.200
   4   0.200   0.200   0.600
   6   0.300   0.100   0.600
   1   0.700   0.100   0.200
   5   0.300   0.600   0.100
```

Author index

Subject index

CPSIA information can be obtained at www.ICGtesting.com
Printed in the USA
BVOW04*1310190813

328920BV00004B/190/A